This book is dedicated to Molly Ferris and her wisdom.

The writing of this book was inspired by the guidance, knowledge and support of my doctor, Nicholas J. Gonzalez. On July 21, 2015, during our final publication stages, Dr. Gonzalez died suddenly. I also dedicate this book to his memory.

Up until his illness, I had only known John as the baker and owner of Baba a Louis Bakery. As an operating room nurse, I see life-changing events. Our community had just barely recovered from the news of another local man with a terminal illness and then came John. It was whispered that he would not survive this cancer. However, he turned out to be our "miracle man." John had his tumor resected and opted out of conventional chemotherapy and radiation. His story reveals the faith he placed in an alternative oncologist who believed that through nutritional therapy, the remaining cancer cells would find a host where they were unable to flourish.

This book is an insight into why certain people may be prone to cancer and invites readers to think differently about this disease. John and his doctor believe that individuals need a certain diet as classified by their autonomic nervous system (ANS). Those that do not adhere to the natural diet for their ANS can open the door to cancer and possibly other illnesses.

John writes candidly about other family members who all had different cancers and describes eating habits and lifestyles that may have contributed to their disease.

There are still so many unanswered questions in science about cancer, its treatment and cause. This book is a must read for those whose lives have been touched by this disease and those in healthcare who diagnose and treat cancer patients. Through our reading, we learn that our diet affects not only our nutritional health, but also our mental health and is possibly a contributing factor in succumbing to other diseases.

John's recovery and survival are a true miracle. By continuing to commit to his nutritional therapy, he will learn more fully what his body needs to remain cancer free.

Leslie Thorsen, R.N.

The author is describing a cancer treatment that he feels is more humane. His commitment to it brought about a transformation. His participation put him in charge of his own healing

A Cancer Treatment is a push to further a cause; a cause that approaches the origin of cancer through the many imbalances in our human existence. He confirms this science by outliving his dismal forecast by 14 Years.

The author has met the criteria of a successful cancer treatment.

Sincerely,
Nicholas J. Gonzalez, M.D., P.C.
June 2015

INTRODUCTION

It was the night before Thanksgiving when one man, MC, began his story by driving himself to the Emergency Room of Valley Hospital.

He could no longer eat. He could no longer sleep. He could no longer deny his sickness. He became the observer of his situation.

What had happened?

The diagnosis, after emergency surgery on Thanksgiving Day and the removal of a malignant tumor, gave him but very little time to live. He was told he had waited too long.

Well, he had waited almost too long.

What if there was a cure for cancer?

What if there was a way around medical systems that push toward compliance without offering any other option?

What if one could find the cancer treatment based on flexibility and sensitivity to individual needs and differences?

Could one avoid the science that sets daily minimum requirements and prescribes one manufactured treatment for all?

As a recovering glutton, MC finds a way to adjust to his ancestral influences, stress, pollution, spectrums of energy and his particular nutritional needs.

This book is about more than just taking supplements and changing one's diet. There is a definite plan of action accompanied by its science.

Cancer does not come and go of its own volition. One needs to participate. Evolution, as well, needs the participation of one's being.

Cancer is not a condemnation. It is perhaps an opportunity. One needs not compromise one's will. Remission is not good enough.

This book is not presented as a step-by-step, how-to manual. It is 13 autonomous chapters addressing cancer from different angles. Along with nutrition, hydration and stress levels, we can add temperamental, psychic, self-determinism, expressionistic, spiritualistic and, especially, the will to diverge from impaired beliefs and science.

If just some of the factors turn off the progression of cancer, why not?

This is but one story. The cancer-free state is accessible to all.

Good luck!

"Until one is committed, there is hesitancy, the chance to draw back, always ineffectiveness. Concerning all acts of initiative (and creation), there is one elementary truth, the ignorance of which kills countless ideas and splendid plans: that the moment one definitely commits one-self, then providence moves too. All sorts of things occur to help one that would never otherwise have occurred. A whole stream of events issues from the decision, raising in one's favor all manner of unforeseen incidents and meetings and material assistance, which no man could have dreamed would have come his way. Whatever you can do or dream you can, begin it. Boldness has genius, power and magic in it. Begin it now."
William Hutchison Murray

1

DR. HEIL MITTEL MANDEL

Home from the hospital and recovering from the blockage removal, MC seemed to have time to reflect. Sharinagar had been of great comfort to him. He was elderly but could still help MC through the worst of it.

He had simply stated to MC, "You either prepare to go, if that is what you want, or you make a decision to go on." This simple statement made a great deal of sense to MC in his post-surgical state of mind. His life was on a balancing point.

MC was leaning toward various alternative theories but Dr. Beauve who was the link to the treatment's direction. Interestingly, it was Dr. Beauve's secretary who would be the catalyst for MC to verbalize his intention.

In doing something other than chemotherapy or radiation, Dr. Beauve, speaking as a friend and acquaintance, offered advice that seemed less cryptic than other doctors' did. Now that MC had declared his intention to seek alternative treatment, Dr. Beauve gave MC his honest opinion: radiation was absolutely useless for colon cancer and chemo treatment for colon cancer had the lowest batting average of all chemo treatments.

Dr. Beauve then gave MC a list of alternative practitioners from all around the country. Finally, he mentioned Dr. Heil Mittel Mandel, who has had an enormous amount of success.

MC had heard Sharinagar talk of the "law of three". Within a short period, two other acquaintances of MC's mentioned they had read of Dr. Mandel or knew of someone helped by him. Oddly, later on, those same people asked MC about Dr. Mandel as if they had no recollection of having mentioned him before. Perhaps they wanted confirmation that the information they had shared was correct and helpful to MC.

MC called all the names on Dr. Beauve's list, finishing with Dr Mandel. He quickly realized that any phone communication between MC and Dr. Mandel's office would be brief and professional. Offering advice over the phone to a non-patient could be troublesome due to a great deal of hostility aimed at Dr Mandel among the cancer research centers. A reference to a Website was the extent of any pre-patient advice from Dr. Mandel's office.

MC was additionally informed that the reversal of his colostomy would be a crucial part in being able to undergo coffee enemas, a critical element of Dr. Mandel's treatment. MC had two months between the colostomy surgery and the stoma reversal, all of which seemed to be laying down the groundwork for his appointment with Dr. Mandel.

As requested by MC, the paperwork arrived from Dr. Mandel's office. The doctor would later refer to this as his screening process. The patient was asked to write about what areas it was thought that help was needed and why he or she wanted to undergo Dr. Mandel's treatment.

MC Screening

I am MC Mac. I was born in 1949. We lived in Vermont. My mother was from abroad and my father is from Vermont. Around the age of three, when living abroad, I developed earaches and was eventually found to have mastoiditis.

I was operated on (1953) in the nick of time, for meningitis could have resulted.

Back in the U.S., when ear problems persisted, I was operated on again. Nothing was found to be wrong, so I was diagnosed with otitis-media and my tonsils were taken out (1955-56.) This would be the last surgery until this recent colon cancer.

Back abroad, (1957-58) my blood was thin. When cut, I reacted somewhat as a hemophiliac - due to too many antibiotics? This passed and, except for a broken leg and a cracked wrist, my health has been good with some ear-throat colds in my teens. From my 20's through my mid-40's I led a fairly pure life style-- no drugs, tobacco or alcohol.

I worked very hard and had some gluttonous tendencies concerning food. I rode out a gall bladder attack after fasting. I would eat as much as a pound of chocolate per day in those days.

A couple of years back I noticed some bloodstained mucous in my stools, eventually resulting in gastric disorders. I did not seek medical attention.

My mother had died at the age of 50. When I began having difficulties previous to my 50th birthday, I started having anxiety about what could happen to me when turning 50. Finally I landed in emergency with an obstruction and had a tumor removed and a temporary colostomy.

I have gotten acquainted with wheat grass juice and diet restrictions. I am now seriously taking them daily. I am also using, since surgery, an Orgone box, which I built when I was 21 years of age. A Hawaiian healer suggested that I eat papaya daily and that I do breathing exercises every day. After surgery, I became aware that I had been living a very resentful, ungrateful and angry type of existence.

My spiritual life has been awakened as well. I am studying "A Course in Miracles" among other disciplines. I am determined to avoid chemotherapy and would like your help in this matter. The reason for wanting an appointment with you is that I feel that my thoughts need correction as well. I am quite convinced that this is the struggle and would like a treatment that supports these efforts.
Yours truly,
MC Mac

During this time, MC had the support of Dr. Beauve whose knowledge of alkaline diets and homeopathic treatments had helped MC along the way. Dr. Beauve was MC's stepping stone to Dr. Mandel.
After post-surgery recovery, MC prepared for stoma-reversing surgery. Dr. Beauve was the man of the hour. It was a stressful winter for MC. He felt severed from his normal function, to say the least. A nurse had

recommended to stoma patients that they use another name for "the bag". MC called his "temp", for temporary. MC needed to heal from the cancer surgery so that reconnection would be possible. Meanwhile, the death prognosis was looming and MC was scurrying to get his will and business paperwork in order.

A letter received from the insurance company stated that only treatments approved by their oncologists were covered by their insurance. A threatening letter arrived from Hilltop Hospital regarding a bill for MC's pre-insurance visit to Dr. Calvaria. Dr. Calvaria had stated back then that the chemo treatment for colon cancer was cheap, only $20 a bag. The doctor had failed to mention there would have also been charges for administering it along with charges for the unwanted advice given. Phone calls were made and letters sent and received in MC's attempt to exclude this pre-insurance charge. A great deal of stress was created as Hilltop continued to insist that MC pay them. The only thing MC remembered from Calvaria's visit was being told he would have brain tumors within six months if chemotherapy were not started immediately.

Soon, a CT scan would show dark spots on the liver and light spots on the left lung. Dr. Beauve would help greatly to interpret the CT scan; however, MC had difficulty reaching him after sending him the scan results. As it turned out, the doctor was ill and had left work for the day. Despite that, an apprehensive MC received a late call from Dr. Beauve the night before his appointment with surgeon Dr. Navaja.
Dr. Beauve reassured MC. "Look", he said, "this doesn't look that bad. They have taken biopsies on these spots, right? So they will look darker anyway. Navaja is going to scare you. Just go ahead with your plan."

The next morning, MC went for his appointment with Dr. Navaja. The doctor said it was a good thing he had not telephoned with this information because he had bad news. "If I were you," he told MC, "I would take a cruise and forget about reversing the colostomy."

Dr. Navaja was a sincere surgeon and an acquaintance of MC's. Despite the previous night's encouragement from Dr. Beauve, Dr. Navaja's news had a strong effect on MC. He felt the colostomy bag fill up rapidly at the suggestion of a cruise. MC thought of the expression, "scaring the crap out of people."

Concerned, the doctor said, "Look, I will do whatever you like. If you want to be reconnected, I will do it."

MC felt grateful. People often criticized Valley Hospital but MC felt the staff was more compassionate there than other hospitals. The more noteworthy Hilltop Hospital, (the source of MC's unwarranted bill), would show less compassion. This was confirmed when MC's friend Rocky was persuaded by contentious doctors to give up any plans for alternative cancer treatment and start chemotherapy. Oftentimes, the more prestigious a hospital, the more persuasion it has over vulnerable patients.

A week before surgery, Sharinagar told MC, "Where there is trust, there is faith. Write that down."
It gave MC something to ponder as he was wheeled to surgery through the halls of Valley Hospital. With the anesthesia taking hold, MC could see that one fears or one trusts; that is an inward decision.

After the reconnection surgery, MC felt challenged by his impatience. He noticed it upon awakening from the surgery. He felt a need to get going - keep the momentum up. This need to be in charge was interfering with the inner calmness to which MC had become accustomed. The second surgery was as difficult as the first. MC knew it was this state of impatience that made it harder. All the problems that had occupied his mind in the past seemed to have disappeared between the cancer removal and the reconnection surgery. Within a few weeks from the reconnection, they began to resurface.

MC had asked his brother Toulouse to come and he arrived to help with the care of their father, Festus. Alarmingly, Festus was discarding goods and property to their stepfamily. Toulouse was also coming to see MC because the news of his health was so dismal.

There was a period of five weeks between the reconnection surgery and the appointment with Dr. Heil Mittel Mandel. It was during this time that the health insurance company announced that there was no coverage for unauthorized treatments with doctors other than those approved by them.

Despite these stressors, MC had a bit more energy and started a protocol of his own. During the weeks before MC's appointment with Dr. Mandel, MC experimented with Vitamin B17, also known as amygdalin or laetrile. A protocol suggested taking 9,000 milligrams of it a day for as many as ten days in a row. On several days, MC took 7,000-8,000 milligrams, but found it hard to function. Dr. Mandel would later confirm that laetrile could be toxic at those levels.

MC began to learn of the political aspects surrounding alternative cancer treatments. The common political theme, from both the right and left wings, seemed to revolve around the fact that cartels or corporations wanted to devise the acceptable treatments for cancer. They could then control the government and insurance agencies to promote these 'acceptable' cancer treatments. MC contacted places that operated laetrile clinics in Mexico. He phoned a doctor there who explained that the procedure was nutritional. The intake of laetrile, along with the oxygenation of blood, was done for about ten days. This seemed to be the crux of the treatment. The doctor affirmed that he was a clinical oncologist and had just returned from Europe, where he had been a guest speaker. He stated that chemotherapy was purely "experimental." He seemed excited and told MC again that he was a clinical oncologist and that he got results. MC mentioned Dr. Mandel. There was a pause in the conversation.

"Yes, we know about him." There was a different tone. For some reason, MC felt he was dealing with a sales pitch but no one was buying. This was not like the reaction of Dr. Beauve and his list of colleagues. They were at least open to the idea of alternative treatments. Nevertheless, MC experimented with the laetrile treatment on his own.

During the post-surgery and pre-Dr. Mandel period, MC was quite thin but never really lost his appetite. He took Essiac tea daily; yet another alternative cancer treatment stamped out of existence. He drank four pounds of wheat grass a week. He enjoyed tofu but would also give that up. Dr. Mandel's treatment allowed no soy, which contains enzyme inhibitors.

There was certainly a lot of information about cancer treatments out there. Remarkably, MC was able to stay on course and found he was more discerning, since, by then, he gave no heed to the medical world. All he did was sort through a maze of alternative treatments. MC's friend Booda always insisted that an oncologist follow or work with him. The negativity thrown off by this alliance was a drain to Booda. In order to preserve his own strength, MC opened up only to those who really wanted to know about the cancer treatment.

The trip to the city for Dr. Heil Mittel Mandel's appointment seemed to be the right decision. There was no second-guessing. The encounters with patients coming out of their appointments and their stories were amazing. Of course, one couldn't know about the cases that didn't work out.

Dr. Mandel came out to greet MC. He asked if Bess, who had accompanied him to New York, could come into the appointment with him. This had been an issue even with visits to an alternative clinic.

Dr. Mandel's reply was, "That is entirely up to you."

This gave MC the awareness that he was in charge of his own treatment.

Dr. Mandel's method was called Enzyme Therapy. This and other supportive methods would have the effect of dissolving invasive growths and foreign proteins in the body. Dr. Heil Mittel Mandel was confident in his results but seemed to be open to all comments and concerns.

He once said, "There is nothing more frightening to a scientist than new information."

He was watchful over the so-called facts about cancer. He seemed to have an intuitive quality and MC could see that his analysis was on the mark with an almost divining quality.

This first visit took two days and blended orthodox and alternative methods guided by two objectives: The dissolving and the discarding of the cancer.

Dr. Heil Mittel Mandel had an expansive overview of the entire process, which included liver and gallbladder cleanses and purges woven into a personal protocol. He told MC that he had an eighty percent chance of success. For some reason, that frightened MC far more than when told four months ago that he had less than a 5% chance of living six months.

Encounters with some other patients were intriguing to MC. He met one of Mandel's long-term patients of nearly twenty years who had outlived her oncologist as well as a dismal diagnosis. What intrigued MC was the conversation with the patient's husband, who was complaining about a health issue he was having. When MC suggested that he do some of the things his wife was doing, the man looked very puzzled. This was not uncommon. MC later realized that Dr. Heil Mittel Mandel's treatment was a well- kept secret. In the time that Dr. Mandel had brought a thousand or so cancer patients to a cancer-free status, hundreds of millions of people had died of cancer. There was even professional jealousy within the alternative medicine world. While Mandel's treatments were copied here and there, including the Laetrile clinics in Mexico, little or no recognition was given him. MC wondered if only those struck with cancer were going to have any interest in the Mandel method.

In the non-cancer world, the most common statement made to MC about the Mandel protocol was, "I could never do what you are doing."

It was a radical change in diet and hygiene and, until one is confronted with terminal cancer, Mandel's treatment seems impractical.

MC started taking more than 100 supplements and enzymes a day. He consumed a quart of carrot juice, which helped fulfill the 70% raw diet that was required.

A liver cleanse, part of the treatment, would actually soften gallstones and then push them out. This liver cleanse was a modification of a three-day apple cleanse that had been prescribed in the early 1900's by American sleeping prophet, Edgar Cayce.

When MC brought the gallstones to Dr. Navaja, he was amazed but quickly said, "These are soft. They cannot be gall stones." As a surgeon who removed gallstones, he was skeptical.

As MC experienced these results, he began to have trust in the method and to realize that the medical world was not all knowing. Perhaps a great deal of healing had to do with actually experiencing these cures, which would result in erasing doubt and skepticism. The liver cleanse was a turning point for MC. From the start of enzyme therapy, MC improved daily.

It is important to note that there had been some improvement with MC before seeing Dr. Mandel. The decision to follow his treatment was itself uplifting. Because MC was trying many things, it would not be accurate to say that a specific treatment was the reason for a cancer-free body. There are also psychological aspects to consider. Scientist Wilhelm Reich had witnessed the disappearance of cancer in people who were able to change their character neuroses. Was there a factor to consider due to the psyche or to spiritual practices?

At one of the visits to Dr. Mandel, Bess said, "I don't see how these methods and simply taking these vitamins can cure cancer."

Without the least bit of defensiveness, Dr. Mandel replied, "You are absolutely right, but we have found that when people follow these protocols, things seem to open up. That is, people start attending church or going to yoga. We do not prescribe any spiritual practice but while following our treatment, people often find their own interests."

Sharinagar often mentions that the teacher and the student find each other. MC came to the realization that the same can be true of a doctor and a patient, and the relationship strengthens them both.

It might be tempting to say here that all MC's problems were solved by Dr. Mandel's protocol; that his body and mind became healthy. MC could be a case in point that one does not need to be enlightened or stress free to become cancer free.

That spring, as people continually commented about how good MC looked, outside problems were developing. MC and his brother Toulouse were watching over their father, Festus, who was doling out gifts to buy some attention. Sections of the Mac homestead had been given away to the kids of Festus's second wife. Perhaps they thought MC would be out of the picture soon and Festus not far behind.

Once again, MC was in the midst of an old and familiar anxiety. However, there was a difference this time. With Sharinagar's help, MC was able to detach himself from these problems and continue to improve with his treatments. As Dr. Heil Mittel Mandel stated, people find the strength and a way to deal with life. MC would always wonder if it was Dr. Mandel's protocol or a shift in mental habits that was more vital.

While talking to Dr. Mandel's secretary, MC noticed that despite lots of phone calls and interest, there were occasional openings in the appointment book, especially in the winter months. One could imagine lines of people extending down the street wanting to see Dr. Mandel.

Though there was no guarantee of being cured, MC wondered how many diagnoses that predict a 5% chance of surviving less than two years have been pronounced on a daily basis. Why isn't Dr. Heil Mittel Mandel's office packed beyond capacity?

One of MC's acquaintances who had heard of Dr. Mandel was interested in talking to MC about him. "It's a shame," said his friend, "that all those negative things are said about Dr. Mandel."

MC said, "I don't see how somebody who has signed and agreed to do this can turn on a doctor."

"Oh, no," she said. "It is other doctors saying these things."

MC realized that in a state of fear and vulnerability one might trust the negative comments made by mainstream doctors more than the positive comments of a patient who has directly overcome cancer thanks to Dr. Mandel. He further realized that people who were curious about Mandel's protocol struggled with mental hurdles.

MC carted around thermoses with a quantity of coffee in them. When the buzz got around that MC did not consume coffee but instead used it for enemas, there were astonished faces and the subject would quickly change. The next obstacle was all the vitamins and enzymes.

Taking hundreds of supplements per day seemed insane, even to MC at times.

Even more absurd was when a chemo or radiation patient, (while taking 18 pills of medicine daily), would say to MC, "I don't think it's good to take so many vitamins."

When following a protocol such as MC's, different types of diets fire up the autonomic nerves, which are also enhanced by different vitamins. Diet is one thing recognized as the most important aspect of a healthy body. Practically everyone agrees that sooner or later we all need to eat better.

MC's protocol was to eat a diet made up of 70% raw food including a quart of freshly juiced vegetables daily, mostly carrots. In his pre-cancer years, MC had learned about juicing and still had the equipment. He consumed the juice of four pounds of wheat grass per week, pre-and-post-reconnection surgery.

When MC asked Dr. Mandel if this was all right, Mandel said, "Wheat grass is the most alkaline food you can eat." Alkaline is what MC needed to eat.

During the first few months of the protocol, MC was drinking a quart of carrot juice a day, which created a slight coloration of the skin due to the carotene. This, along with the sun, gave MC a bronze complexion that was the envy of some. Some would even drink carrot juice just for that look. When MC went to Dr. Navaja for a checkup, the doctor began to inquire about this coloration.

When he saw the color in the non-tanned area, he said, "This could be dangerous to the liver. We need to give you a blood test."

Since MC needed a blood test for Dr. Mandel, it was convenient to have it ordered by MC's surgeon. MC noticed that no news about blood tests was usually good news.

When he brought this up at his next visit to Dr. Mandel, the doctor said, "At the beat of an eyelash, those folks will put people through near death experiences with chemo but get all excited about carotene coloration."

MC began to realize that the science about cancer is not at all on the same page. On one side, there is an established field trying to prolong life a few months by throwing all the inventions of the atomic age at it. Protons, accelerated in a compound the size of a football stadium and costing hundreds of millions of dollars, try to pinpoint a tumor without burning the whole patient. On the other side was this less well-established treatment with its vitamins, raw food supplements and the use of pancreatic enzymes. The intake of these enzymes must be monitored so the tumor doesn't dissolve faster than the body can discard them. Tumors that dissolve too quickly sometimes result in kidney failure, which stresses the importance of timing and cleansings.

At the five-year anniversary of MC's initial meeting with Dr. Mandel, the importance of continuing a pure diet was stressed, especially as the threat of death receded.

At the end of the visit, MC told Dr. Mandel, "Nobody believed this can work. It is too simple."
Dr. Mandel nodded in agreement.

MC then said, "People do choose their cancers."

"What do you mean?" asked the doctor.

"Toulouse's cancer was in the brain, Popsy's in the lungs. If cancer was not somewhat selective, the symptoms would all be the same and would take the same path through the body's organs."

Dr. Mandel replied, "Oh, I see what you mean."

MC had taken a leap of faith to undergo the Mandel method. He had never even watched his videos or listened to his tapes. Unlike his friend Booda, who was convinced by explanations of how it all worked, MC just trusted that this was the way to go.

What would really convince MC of the effectiveness of Dr. Mandel's protocol was coming across his own report written by Dr. Mandel about four years into his treatment. During the year of this report, about 150,000 new cases of colon cancer were diagnosed and about 60,000 deaths had occurred. Only lung cancer claims more lives. The rate of death in colon cancer has stabilized perhaps due to public awareness campaigns emphasizing regular colonoscopies and the removal of colonic polyps.

MC's Report

Colon cancer has been linked to environmental factors, especially diet. A high intake of animal fat and presumably the conversion of saturated fatty acids into carcinogenic compounds in the colon increase risk. High serum cholesterol, obesity and colon cancer have a correlation. The intake of fiber in those over 50 does not seem to have the "deterrent effect once claimed. This age group, (MC's own), is most vulnerable to colon cancer.

Let us divide colon cancer into four stages, which are based on the depth of tumor penetration in the bowel wall.

Stage 1: Cancer is limited to the superficial layers of the colon with no invasion of underlying tissues.

Stage 2: Indicates the tumor has invaded the bowel walls but not into the regional lymph nodes.

Stage 3: Signifies the disease has spread into local lymph nodes.

Stage 4: The worst; the disease has metastasized to distant organs such as the liver or lungs.

90% of patients with Stage 1 colon cancer will live five years, while about 5% with stage 4 last that long.

Patient MC a Seven Year Survivor

Patient MC is a 60-year-old man with a family history of cancer. His mother had brain cancer and his brother, Toulouse, succumbed to renal cancer just recently. Also, two uncles had cancer.

MC had generally been in very good health when, beginning in 2000, he noticed a change in his bowel habits, including increased mucus in his stools, chronic indigestion, bloating and what he described as gas pains. Hoping for some relief, he adopted a whole food, vegetarian way of eating. But, through time, his symptoms only worsened.

In mid 2001, he first noticed intermittent bright red blood in his stools. Some months later, in October 2001, he developed symptoms consistent with a bowel obstruction including severe pain, bloating, abdominal distention and an inability to move his bowels. When the symptoms resolved after several hours, he chose not to seek medical attention.

Several weeks later, in November 2001, the symptoms returned with a vengeance. He hoped once again to ride out the crisis, but over a three-day period the pain, bloating and distention worsened and brought him to the local emergency room.

A barium enema revealed an "apple core" lesion in the sigmoid colon, indicating a tumor. A subsequent sigmoidoscopy revealed a complete obstruction of the colon and the patient underwent emergency laparotomy, resection of the sigmoid colon along with the tumor and the placement of a temporary colostomy.

The surgeon (Navaja) also discovered as his operative notes report, "palpable nodules in the liver, which I felt to be more cystic than solid, but there were a couple of studs that were solid." He removed one of the liver lesions for evaluation.

The pathologist's summary describes a large colon tumor, but doesn't give exact dimensions, though it states, "The mass locally grossly appears to extend to the underlying adipose tissue" and defines the tumor as "moderately differentiated adenocarcinoma, extending through the bowel wall, and present on the serosal surface." Though cancer had infiltrated two of the lymph nodes examined, the liver tissue seemed most consistent with a benign hemangioma.

Postoperatively, a CEA (Carcino Embryonic Antigen) test, a tumor marker for colon cancer, came back elevated at 5.1 (with normal being less than 3) an indication of remaining malignant activity. No CEA had been done before surgery so there were no results for comparison.

MC did subsequently meet with an oncologist (Calvaria) who suspected that the tumor had invaded the liver, despite the negative biopsy. He insisted chemotherapy needed to begin quickly, but upon questioning, admitted that if the cancer had indeed spread, treatment would do little.

MC, with a strong interest in alternative medicine, decided to refuse conventional treatment and instead, began self-medicating with a variety of nutritional supplements. After learning about our work from a caregiver friend, (Dr. Beauve), he chose to proceed with our treatment. He contacted our office in early January, 2002, but we suggested that he come in only after reversal of his colostomy.

Since the patient had been rushed into surgery in crisis (Thanksgiving Day) from an obstruction, no preoperative CT scan had been done. Finally, in mid January, his doctors pushed for a CT scan, which revealed evidence of multiple metastatic lesions on the liver as the official report describes:

"Unfortunately, within the liver there are numerous small hypo enhancing lesions, some of these are very hypo enhancing to the point where one might consider them cysts, but others are of a more intermediate density. 5mm-thick slices were obtained to increase the sensitivity. The largest of these lesions is only about 1 X 1.5 cm. There are symptoms for metastatic disease."

The radiologist also noted 'very minimal sub-pleural densities seen at the mid left lung field' which he felt 'should be rechecked within several months.' In his summary, he reports that 'I suppose that the liver findings increase suspicions of the left lower lobe findings; however my feeling is that the lung changes will prove to be benign.'

Quite likely, based on the CT findings, cancer had spread into the liver and possibly to the lungs. The negative liver biopsy, the patient was told, might only indicate that the liver contained both benign and malignant nodules, as the CT scan seemed to show.

In late January 2002, MC returned to surgery for reversal of the colostomy and lysis of adhesions that had formed since the first operation. During the procedure, unfortunately, none of the liver lesions was biopsied.

When MC was first seen in my office in mid March 2002 he seemed enthusiastic about the therapy and subsequently followed the regimen faithfully. Today, more than seven years on treatment and 7+ years from his original diagnosis, he remains fully compliant and enjoys excellent health.

Over the years that he has been my patient, MC has chosen not to undergo any further CT scans, a decision I have respected. He says no matter what the scans show he wouldn't agree to chemotherapy nor would he change his treatment. He doesn't want the radiation exposure (which is significant), the worry or the expense. So I have no idea what has happened to the liver or its lesions. I only know the patient is alive and well.

Even if we disregard the CT liver findings for a moment, a number of salient signs point toward a dismal prognosis. The literature reports that patients who initially present with an *obstructing lesion have a far worse prognosis than those who don't, even if the disease is otherwise localized. In MC's case, the fact that the tumor had already invaded through the bowel wall and infiltrated into two lymph nodes signaled future trouble. The carcino embryonic antigen (CEA) level after surgery, though only mildly elevated, "nonetheless also warned of a future recurrence - regardless of what may have been going on in the liver. MC's elevated postoperative CEA served as an even more worrisome prognosticator. Yet, if we accept the expert radiologist's conclusion that cancer had infiltrated the liver, the prognosis turns dire.*

Synchronous liver metastases (meaning liver metastases occurring at the time of the original diagnosis of colon cancer) has a survival range of 4.2 to 8.7 months, let us say six months with aggressive chemotherapy. MC with multiple malignant appearing lesions on CT scan not only has far outlived the predicted lifespan but has successfully avoided the toxic treatments his oncologist insisted seven years ago needed to be done.

2

ISIS: A REFLECTION

Isis was connected to MC through marriage. She was fortunate to have the means to travel often and remain unemployed. Because of this, she was able to live in different parts of the world.

Outspoken, Isis had a tendency to take charge, especially when she observed a lack of direction. Occasionally regal, Isis was an appropriate name.

Isis ended up announcing that she had leukemia. Her first chemo treatment would take place on the same day that MC walked into the emergency room with a complete blockage of the colon. It appeared that two members of an extended family fell on the same day. Isis and MC were dealing with their separate issues long before they surfaced. Now there was a common bond between them. This helped MC understand how the outside world and family circles react to the news of cancer. To most, cancer seems to be a curse that strikes quickly and is interpreted by the multitudes as the beginning of the end.

Isis quickly tried to establish business as usual and take everything in stride. She had a lot to say about the social aspect of leukemia treatment and about how interesting the science behind the treatment was. For her it was a new identity, a new quest on surviving cancer, a war against cancer.

Isis had the advantage of being unscathed by surgery. After a series of chemotherapy, her blood count numbers were such that further treatment was postponed. MC on the other hand was remembering encounters with people who had dealt with colon cancer. One guy had just finished taking care of his father who had died of colon cancer in its allotted time.

"My father lasted as long as they usually do with colon cancer -- a couple of years".

MC was rejecting the odds and the timetable of simply biding the ordeal with no experience or countermeasures. He would soon aspire toward a new approach.

Isis had complete confidence in her doctors and seemed optimistic even though things would eventually go downhill.

Meanwhile, MC was about as pessimistic as one could be. He was uninsured and in the hospital with a stoma and open incision that needed to be dressed two or three times a day.

Dr. Navaja had a sad look as he said, "There's good news and bad news. The good news is those spots on your liver tested negative. The bad news is two of the nine lymph nodes outside the piece of intestine that was removed were found to be cancerous."
He was gently telling MC that his cancer fell somewhere between category three and category four.

To put it mildly, MC was not all that thrilled about his cancer or its science.

As he reflected after his surgery, MC spent most of his time staring at the ceiling and noticing its defects. He remembered that Sharinagar had told him months before that he looked more miserable than ever. He was cognitive that something was different. His way of life and rounds of negative thoughts that continually plagued him, (losing the family farm, a former business partner making off with the company name resulting in direct competition with MC, as well as dealing with dishonest former employees), was never resolved or emotionally worked out by MC. He was resigned to the fact that his fate was the way of the world. Now all of this began to pale compared to MC's present situation. He knew he needed to remove these thoughts, as there was no room in his mind but to deal with the problem at hand.

At the onset, MC had taken this diagnosis as a sign of having flunked out. It humbled him as he grappled with the idea of having allowed this cancer in the first place. He came to feel that he would need to take responsibility for this cancer - to recognize it as the opportunity of a lifetime - to outgrow and soar above it.

Isis and her husband began receiving messages that MC was drifting away from the idea of chemo treatment. It became a crusade of theirs to try to change his mind. With no email, MC received a $20 express mail envelope from Isis containing a single-page letter pleading that he be reasonable and thinks of his son and the rest of the family.

Since Isis's chemo treatments had been postponed, she would get some time off. By summer, Isis and MC came together with other family members at a summer house.

Unlike Isis, MC dreaded having to explain so much about the process of his cancer treatment to everyone. One hundred or so pills a day were taken, raw vegetable juice was drank to complement a 70% raw diet and coffee enemas were an everyday occurrence. These are things that have to be planned around the activities of others. Isis had the knack for dumping MC's coffee out, wondering what a pot full of coffee was doing on during the latter part of the day. Fortunately, it turned out that most people weren't interested in MC's daily routine. They began to notice that he was doing well and looking good. The carotene from the carrot juice combined with the sun and salt water was giving MC what they called the "Bronze Man" look.

During this time, Isis commented on the success of both systems, "You're doing well and I'm doing well. We're both treating our cancer differently but with good results."
After his brother Toulouse's death, MC learned that he and Isis had different forms of cancer: parasympathetic dominant and sympathetic dominant.

As MC's cancer had been manifesting, his financial worries had been increasing. When talking to Isis about cancer, he mentioned that its three principle factors are dehydration, malnutrition and stress.
Isis agreed, "Oh yes, especially stress, that is the big one."

MC then realized that one's thoughts and stressors are independent of wealth and leisure.

In her pre-leukemia days, Isis could be challenging and manipulative. One could imagine a cause and effect in this behavior. After the diagnosis, she mellowed out. She never seemed to have much spiritual faith. Instead, she relied on her body's strength, an innate constitution and banking on inherited good health. MC was conscious of this way of thinking.

Isis began to have difficulties. Sunspots appeared on her scalp. When they were burned off, oozing areas were left behind. She bruised easily, one of the characteristic symptoms of leukemia. While skiing abroad that winter, her spleen began to enlarge. The spleen is part of the lymphatic system and leukemia is cancer of the lymphatic system. Her spleen was surgically removed while in Europe.

MC and his son visited Isis that spring. The visit occurred during a break in her chemo treatment and things were not going well for Isis. She had become curious about what MC was doing. He had never really described the cancer treatment to Isis. She asked MC to tell her what it involved. MC told her of the daily coffee enemas and of the nine bags of pills ingested daily. These included six bags of pork pancreatic enzyme, bromelain and papain, one bag taken every four hours. Taken with meals were the three remaining bags containing vitamins, minerals and animal tissues.

Isis seemed interested in what his treatment involved. Without any bias or sales pitch, MC continued describing his protocol of purges, cleanses, vegetable juices and his required diet.

Suddenly, Isis turned her head away and said, "No offense, but I'd rather die than do what you're doing."

Her declaration seemed to give her the resolve to go on. MC never thought of his protocol as a deprivation or an imposition.

Isis was forcing herself to eat, which usually resulted in vomiting. She was losing a significant amount of weight. Her friends had located a doctor that would make house calls. The doctor came that evening and Isis was pleased to see him. A pleasant man, he examined Isis in her bedroom. Sounding energized, she thanked him profusely as he left.

Isis entered the room where MC, his son and other family members were gathered. She announced, "This doctor has told me that all my symptoms are entirely due to the chemo treatment and not due to myself."

MC anticipated that she was going to proclaim the end of her chemo treatment but the opposite logic occurred. As long as the chemo, and not she, was the problem, Isis gained courage to continue the treatment. Perhaps already too late to alter the events, Isis needed to reaffirm her choice. Too much had been invested to pull out now.

On another occasion and with tears in her eyes, Isis told MC how difficult it was to get a smile from the poker-faced surgeon that had removed her spleen. Now she boasted that she had opened him up and had a pleasant conversation with him. In conclusion, the doctor joked with her saying that the spleen removal had resurrected her. This small consolation prize, appealing for some human warmth from those who were treating her, was an important victory that comforted her. Isis's pride, the prestige of the doctors and the rapport between them seemed to offset the lack of results.

As MC and his son were leaving the next day, Isis said that she felt a little better. "As soon as I feel a little better, it is time for my next chemo treatment."

Isis's husband reported that she never actually completed a prescribed treatment. Up to then, they had all been cut short. Despite, (or because of), the chemotherapy, two months later Isis's blood counts worsened. The hospital informed her husband that the chemotherapy would be ended. It had intermittently been prescribed to her for three years, the cost of which would have paid for at least twenty years of MC's cancer treatment.

At one point, there had been a plan for Isis to try some different treatment. Maybe it was suggested to give her hope. However, by grace, or by lack of ethics, the hospital backed out and she was left in bad shape.

Isis had been given a certain time frame as to how long she would survive. She and the family returned to the U.S. where things degenerated more quickly than had been forecast.

She was overheard saying, "Why did this happen to me?" She passed away in a hospital.

Isis had been in her mid-sixties. With friends from all over the world, there was an enormous gathering at her funeral. Afterwards, an acquaintance came to visit MC and they talked about Isis. At first, MC thought his friend had a similar view on the detrimental effects of chemotherapy. Suddenly she clenched her fist and said, "It's too bad she did not get the chance to try the different treatment. You know, abroad they might not have all the know-how. If she could only have been given one more dose, maybe the cancer would have been knocked right out of her!"
Once again, despite the outcome, MC marveled at the blind trust people have in this harsh monopolist science.

A Cancer Treatment

3

BOODA: BUMPS IN THE ROAD: THERE IS SUCH A THING AS WAITING TOO LONG

Booda always sat close to the front during Sharinagar's sharings.

Sharinigar would usually acknowledge him in a mock adversarial manner. During a lighter moment at one of the conferences, Sharinagar was speaking of the commercialization of spiritual studies and how one can never sell such things.

Suddenly, while pointing at Booda, he said, "If I really wanted to make money, I would take him, take off his clothes, put him in the back of a pickup and go around the country selling him as the Buddha." There was an absolute roar in the session. Booda got his name.

Booda was a former college football tackle and MC had met at a retreat. Booda endured the death of his wife, who had undergone traditional chemotherapy as well as a smorgasbord of alternative treatments. In the end, Booda and his wife were heading to Mexico to try more things. By then, the whites of her eyes were a florescent yellow and things were so far along that no treatments were possible, traditional or otherwise.

MC remembers Booda saying at his wife's funeral, "One should stick with one thing and not mix it up."

After Sharinagar passed on, there was still a friendship among the retreat people, who continued to meet but less frequently.

Booda developed cancer of the esophagus with ominous CT scan spots. This put him into the category of advanced-stage inoperable cancer. Esophageal cancer is considered terminal within the year and there is little hope that it will respond to chemotherapy. A mutual friend suggested to Booda that he contact MC.

MC noticed a strange phenomenon - even though it was known that MC had dealt with cancer, very few people heard the details unless cancer popped up somewhere in their own lives or families. When someone survives, the thinking is that it must not have been too serious in the first place. MC never wanted to make public his dismal forecasts. In fact, he himself never acknowledged these forecasts. Booda kept his diagnosis under cover for a while, as well. Finally, he contacted MC with the news and told him that he was very interested in learning more about MC's treatment.

During Sharinagar's conferences, Booda and MC had occasionally shared a room and Booda recalled some of the accoutrements of MC's treatment: the juicers, coffee machines and bags of pills. After seeing a DVD of Dr. Mandel's, he decided to pursue the method more thoroughly. When Booda put his mind to something, there was intent. He made a decision. His hope was to see if the hospital oncologists would work with him by providing approval to follow Dr. Mandel's cancer treatment. To MC this seemed a turbulent alliance. Even though the oncologists' view made clear there was little to offer for even prolonging Booda's life, the mention of an alternative treatment triggered a militant stance from the oncologists.

Defensiveness is something that strikes us all. When a professional starts getting defensive while your wellbeing is at stake, you see how little regard there is for the human body. They would rather terminate life in favor of their beliefs than acknowledge another way or truth. How many people undergo treatment to please other people, spouses, parents or not to offend relatives and caregivers? Booda felt that he needed to explain Dr. Mandel's methods to his loved ones and dealt with the negative comments from the medical professionals regarding these methods.

When Booda gave the information to the oncologist, she glanced at it for five seconds and said, "This will not work. You are wasting your time."

Another doctor said, "They are taking your money."

These observations were coming from a very costly industry that had nothing to offer Booda at that point.

MC was well aware that when you opt for alternative treatment, you have to concentrate on that and nothing else. MC often felt that the cancer patient in the oncology office is a bit like a mouse trapped by a cat. Despite its will to survive, the mouse's chance of escape diminishes if it becomes injured. It will often become disoriented and confused. The more its tenuous path twists, the more likely it is that the vulnerable mouse ends up running right back to the cat, the future of its survival now in the cat's hands.

Booda made some immediate improvements. He lost 60 pounds in about three months. He had dealt with high blood pressure and diabetes his entire adult life but in a short time, his sugars were normal and he was almost off his blood pressure medicine.

As Booda had asked the oncologists to work with him, they wanted to follow what was going on. In three weeks, the CT scan showed that the esophagus had stabilized, but there were spots on his liver. Plenty of good liver left, they affirmed, but this sort of interaction was always stressful and they could never leave out small bits of negativity for Booda to take home.

Booda commented that Dr. Mandel said, "MC was as bad off as you are." This gave Booda encouragement.

The treatment seemed to be taking its course. There were little bumps, as Booda called them but he had a way of painting a positive picture. He never offered MC detailed information but made it clear that he was grateful to MC for having the opportunity to undergo the treatment. Booda's stories about all the possible mishaps of coffee machines, enema bags, juicers and pill bags created uncontrollable laughter. He spoke of the reaction of a world oblivious to treatment that does not fit any treadmill.

MC called Booda one day and discovered that he was in the hospital with a high white blood cell count and fever peaks once or twice a day. While talking to MC, Booda was jovial and funny. Suddenly, he became serious and told MC, "I realize now that there was cooperation between my body and this cancer. It went relatively unnoticed because my body did not perceive a foreign object. I now see that the body wants it out and with the dissolving of some of this cancer, it has become like gangrene".

Booda then said it was more difficult to follow the protocol in the hospital but that some of the nurses were interested in his procedures and helpful to him.

After the phone call, MC wondered how Booda was really doing.

When MC called three weeks later, complications were on the rise. Booda was tired and not as clear or lighthearted as usual. Knowing that Booda was not well, MC began calling friends and relatives. A friend told MC that the cancer had been diagnosed in the bones, (something Booda never told MC), and that at this stage, his tailbone and ribs were breaking, creating extreme pain. "Booda isn't for long."

MC called Dr. Mandel and explained that he had a headache similar to the one that he had when his brother Toulouse passed on.

"Doctor, do you think Booda is dying?"

"Yes, I'm in contact with his wife and parents."

"I feel very mortal, doctor."

"Don't you go there," Dr. Mandel advised. "Your friend Booda's illness was very advanced. We almost turned him down for treatment but because he was your friend, so willing and such a nice guy, we agreed to go ahead. He was very well aware of this, but when you are in and out of hospital while taking antibiotics and pain medications, it is very difficult to follow the protocol or any orderly treatment."

MC told Dr. Mandel that at first Booda seemed to make enormous progress. The doctor agreed and added that had Booda begun treatment a few months earlier, the results may have been as good as those MC had experienced.

In his stories, Booda had always left out the big bumps. Now, at the final stages in the hospital, Booda was made to feel like a dissident. Even though the treatments at that point would have been palliative, the doctors' scorn was plainly evident.

Booda had been aware of the dismissive attitude of the oncologist when told, "What you are doing is not working."

Booda and his family were silenced when a physician declared to Booda, "We need to get a resuscitative unit declaration", which meant that Booda needed to decide if he wanted CPR in the event he stopped breathing. The doctor continued, "All patients on this floor need to make this declaration."

Booda replied, "Do all you can to revive me. My children have the authority to shut things down. Treat me as you would any other patient on this floor."

In a dismissive tone the physician said, "There is no other case like this on the floor."

The family, who seem always to be aware of what is going on long before the professional does, was bothered by the doctors' need to get in the last word.

Booda's pain was on the increase. It was suggested that treating the coccyx with radiation might alleviate some of the pain. With Dr. Mandel's, blessing Booda was willing to try it.

Now that the hospital was fully involved in treatment, the attitude of the medical unit took a turn as they prepared for the radiation procedure. Booda was agreeing to a procedure that he did not believe in just to alleviate the pain.

MC saw the anguish of Booda's situation inside the hospital. It was like someone going into a restaurant but insisting on eating a packed lunch from somewhere else.

Booda started failing. It was almost impossible for him to move him. It was assumed that ribs, a hip and the tailbone were broken due to the cancer. After a blood test, it was concluded that both his liver and kidneys were failing.

A physician announced these test results and stated, "He is no longer a candidate for radiation."

A short time later Booda said, "They don't decide when I go."

A few days later, Booda died in his childhood home.

MC began to get information about Booda. Three years prior, when he met his second wife, Booda knew he needed to alter his diet. This cancer took a while to get started but Booda knew he had brought on his illness after years of abuse and poor diet. Booda had told MC that he had high blood pressure, diabetes and sleep apnea for most of his adult life.

During the twenty years of Sharinagar's lectures, Booda had always sat in the front row. How many times had MC and Booda heard the words: "The truth once heard and not heeded will turn to poison"?

At the very end, Booda again thanked MC for the opportunity to do the cancer treatment. He had made the decision and felt results, however temporary.

Any compassion Booda had hoped to feel from the hospital oncologists went unrealized. Any words of encouragement about his decision to treat his body in a way that honored his instincts would have been welcome but, not surprisingly, never offered. Booda felt betrayed by the medical industry's regimented ways. When it is decided there is nothing more the industry professionals can do to help, a patient is seen merely as a specimen to be treated in the same way as the next.

At Booda's wake, MC came to realize that waiting and postponing could have consequences beyond the possibilities of the cancer treatment. This was a sobering experience to MC's enthusiasm about his cancer treatment.

While driving home after Booda's funeral, MC listened to a live radio broadcast of a benefit drive to raise money in the hope of eradicating cancer. The drive had grown to be an annual, multi-million dollar event. Afflicted and depleted bald children were paraded about to help raise funds before the baseball game. An insurance company was in charge of collecting the donations. One of the benefactors was the hospital where Booda had struggled so.

4
ROCKY: A COMPROMISE

Rocky was a very social being. He was in the culinary field and always wanted to have a restaurant named "The Maine Potato".

He was loud, entertaining and loved jokes. People would throw any joke at him and he would usually respond with a moan and top the joke. He could participate in any argument with an adversarial stance that humored rather than offended. People working with him would say that he was expressive and to the point. He had an enormous number of friends in all walks of life. MC could state a problem and Rocky would be right on the same page with him. Rocky had been to Vietnam but was also of the sixties. He worked hard and eventually got into a restaurant partnership.

Rocky would often drop in to pick up goods at MC's business. One day, MC gave Rocky something to read and noticed that his hands were shaking. Rocky had the look of a tough man but MC thought to himself, "Rocky's got the shakes. Something is wrong with him.

Shortly afterwards, Rocky was diagnosed with melanoma on the lower back. The news rocked the county. As MC had done, Rocky got on the Internet and announced that he had found many treatments - from mushrooms to teas. Rocky was aware that MC had come through a bad dilemma and they talked in confidence. MC gave Rocky the information on Dr. Heil Mittel Mandel but in the background, there were plans for surgical removal.
After Rocky's surgery, the cancer was discovered going up his back.

In a situation such as this, there seems to be a line drawn between a slow alternative method, with its "wishful thinking", and the orthodox mentality of seek, burn and destroy. This whole proceeding was moving quite fast and, once again, there was an urgency to start treatment immediately.

Knowing that radiation and chemotherapy takes a lot out of you, Rocky was determined to combine alternative and mainstream treatments. He said the doctors seemed willing to work in tandem with other methods. This reminded MC of a conversation he had with Dr. Mandel's secretary about an article on treating cancer with a combination of chemo and alternative treatments. He noted the patronizing and mocking attitude of the professionals involved. The patient would listen to ambient music, burn incense and place quartz crystals around the hospital room during the chemo treatment. MC knew how things happened in hospitals. Before long, all artifacts are put back in their place. In other words, turn off the noise, put out the smoke and put the rocks in the closet.

Despite his hopes, the doctors soon convinced Rocky to drop his alternative treatment plan and begin radiation treatment.

A few months later, negative news brought on the proposal for chemotherapy. The news about Rocky's treatment was widespread and, still in character, he struggled to make it look easy.

Shortly afterwards he told MC that the chemo seemed to be driving the cancer everywhere. There was indication that some of the organs, including the liver, were showing signs of cancer. Then, suddenly, the treatment stopped. After the rush to start Rocky's radiation and chemo treatments, time seemed to drag on after the treatment ended. His body seemed to spring back a little but the problems persisted -- nothing had been corrected.

Around the Christmas holidays, Rocky dropped in to see MC. His ears looked thick and dark, almost black at the edges like something seen in a space odyssey movie. His complexion was jaundiced and gray. He wanted to talk about what MC had experienced.

He kept saying, "MC we have to talk about your methods."

However, since it was a busy time of year, Rocky didn't want to talk then. He died less than a month after that meeting without ever summoning the energy to talk or debate whether he had made the right decision for his treatment. Rocky told MC that he had doubted the use of chemo and radiation and regretted not having tried Dr. Mandel's treatment.

In his early fifties, Rocky died a mere nine months after his cancer diagnosis. MC always wondered if he should have gone to Rocky sooner. Sadly, he passed away on the day MC had planned to see him.

His funeral was an enormous party with food and music. Several people spoke at the service, among them cancer survivors, (or should one say survivors of their cancer treatment?), who spoke of the fight against this insidious disease and how Rocky had succumbed to it.

5

RADIO TALK SHOW ON BREAST CANCER

The frequency of breast cancer is at epidemic proportions in the United States and Europe and now China is catching up. Even third-world cultures are beginning to show signs of an increase in breast cancer cases, especially in urban areas.

While listening to a radio program, MC heard a guest promoting the need for mammogram imagery machines and advocating for prompt treatment in treatment centers all over the world. In conclusion, the guest lightly touched on the possibility of the adaptation of the Western diet being among the reasons for the increase of breast cancer in these third-world countries. Although the radio show host returned to the issue of diet on several occasions, the guest failed to elaborate. Despite her own mention of the problem, her take on remedying the problem leaned towards the belief that mainstream Western treatments was the only way to go. She believed that more facilities were needed for screening and that chemotherapy, radiation and surgery would nip the problem in the bud.

The radio host took a call from an African woman who had dealt with breast cancer. She began her comments by lamenting that although breasts are sexualized in the West their primary use, for feeding infants, was frowned upon among an increasingly puritanical public. Everyone agreed with that observation. The guest speaker went on about the tenacity of breast cancer and mentioned how the more serious cases are primarily among African-American women. This opinion usually leads to a conversation about genetics, which is often the fall back reason for cancer when there is a lack of understanding about the disease. After all, more research is needed in that area, as well.

A caller pointed out that socioeconomic factors might play a role in the increase in these virulent cases of cancer in black women. However, she was unable to finish her thoughts about the role of diet as the guest speaker stuck to her statistics.

Again, the host asked his guest for her explanation for the increase in the disease. This time he directly encouraged her to comment on the role of Western lifestyle and diet.

The guest said, "Look, more and more of the world's population is seeking an easier way of life."
She went on to say that part of this massive change is the adaptation of a diet that mimics the West's with the focus shifting to faster, processed foods. The office-work lifestyle leads to the desire for convenient and quick remedies to daily nutritional needs. She suggested that technology and medicine needed to adapt to this trend. Women should be encouraged to have regular mammograms and, when

needed, get the proper care and undergo immediate treatments. Not one example of Western diet was talked about; not one food was mentioned; and not a single nutritional **concept** was discussed. Maybe this would offend vested interests in the Western-style economy?

In another radio program, a doctor was stating that chemotherapy treatments for particular cancers were problematic because the chemo seemed to be attacking healthy cells while cancer cells were resisting. Although this doctor was probably being quite courageous in stating these facts, he excluded any other approach to remedying the existence of this problem except to say that a new chemical needed to be found. This seemed arrogant, as he had already openly admitted that more harm than good was being done to the patient undergoing chemotherapy.

As he listened, MC wondered how long Dr. Mandel would last if he came on the air and announced that *his* treatment was doing more harm than good?

For MC, the cancer industry has an unspoken and absolute rule. Their "failed state" is content with measuring the shrinkage of tumors and trumpeting the survival rates to a disease they believe incurable. Their fragmented findings are interpreted through a collective using such terms and sound bites as "on the verge of" or "experts say", with no particular commitment to finding a cure.

Their moralist arguments to go on with their devastating treatment is justified by obedient and dutiful patients fulfilling their obligatory task, hoping that herd mentality and mass belief might lead to a better prognosis.

6

TOULOUSE: A STUDY IN THE
AUTONOMIC NERVOUS SYSTEM

I f cancer is predetermined or inherited genetically, how can it be so dependent on diet, a particular temperament, (the autonomic nervous system), and environment in order to manifest itself in a particular body location?

MC would experience the cancer phenomenon in himself as well as his whole family. It was not easy pinpointing the inherent links to cancer in MC's family. His father, mother and brother represented the three main metabolic types in the autonomic nervous system.

Until the age of 53, MC lived unaware of his true nature or his ideal diet. This awareness came to light because of his diagnosed terminal colon cancer. The awareness grew when helping others with their cancers. It was a confirmation to MC that something quite simple determined cancer and cancer types that was overlooked by nutrition, physiology and oncology. MC's brother Toulouse supplied MC with the encouragement to comprehend the role of these autonomic nervous systems and their relationship to cancer.

The autonomic nervous system, (ANS), is the nerve network that regulates all aspects of the metabolism that do not require any conscious input.
The human species is divided into three genetically-determined groups:
- Sympathetic dominant
- Parasympathetic dominant
- Balanced metabolizer

From "One Man Alone: An Investigation of Nutrition, Cancer, and William Donald Kelley", by Nicholas J. Gonzales, M.D.

"These groups would have different temperaments, require different diets and get different types of cancer.
Kelley recognizes gradations of "wrong" eating. Most of us follow the right diet for our type some of the time, and alternate "wholesome" with nutrient depleted foods
Overall, Dr. Kelley insists the more we stray from the ideal diet, in terms of both quality and content, the further we move from autonomic equilibrium and metabolic efficiency. If the problems aren't corrected with appropriate diet and proper nutrition, overt disease inevitably develops.
"Dr. Kelley actually designed a very sophisticated graph, to illustrate his concepts of autonomous imbalance and metabolic inefficiency. In his scheme, the horizontal axis represents autonomic function, with the far left point indicating extreme sympathetic dominant, the far right,

extreme parasympathetic dominance, and the middle point on the axis, balanced autonomic activity.

"The vertical axis, on the other hand, charts overall efficiency, with the highest point correlating with optimal autonomic balance and ideal physiologic function, the lowest point, catastrophic decline in both autonomic divisions and in the various tissues, organ, and glands. Patients who follow the "right" diet for their type, (plant-based for sympathetics, largely meat for parasympathetics, and varied for the balanced), but who consume nutrient-depleted, refined, processed, chemical-laden food usually end up at the bottom of the efficiency axis in the Kelley scheme, with both autonomic branches hobbled, and all tissues, organs, and glands functioning poorly. Whether innately sympathetic, parasympathetic, or balanced, all tend to converge at the same low vertical point, plagued with problems such as adrenal failure, gonadal inefficiency, hypothyroidism, and poor sleep. Often, such patients bounce from doctor to doctor, seeking help for their vague but relentless symptoms and chronic ill health. Those who consume the wrong diet for their type inevitably move to either of the extremes on the horizontal axis. In such cases, Dr. Kelley associates very specific syndromes and illnesses with heightened sympathetic or parasympathetic dominance. For example, a sympathetic on a meat diet might at first experience anxiety, hyperactivity, restlessness, increased mood swings, and insomnia, as his sympathetic system becomes chronically more active. His--or her--powers of concentration might decline.

"As the leftward drift continues, sympathetics can develop chronic problems such as gastro-esophageal reflux, colitis, or insulin dependent diabetes, due to the inhibitory effect of the sympathetic nerves on the digestive organs and endocrine pancreas. These patients commonly present with high blood pressure, due to the vasoconstrictive action of the sympathetic neurotransmitters. In extreme imbalance, sympathetic dominants can fall prey to severe, even life-threatening, illness, such as congestive heart failure and devastating strokes, as well as psychiatric problems such as hypomania or schizophrenia.

"A genetic parasympathetic adhering to a predominantly plant-based diet, as he or she moves further into parasympathetic dominance, might initially require more sleep but never feel rested, and report chronic low grade depression, persistent allergies and hay fever, all due to increased parasympathetic firing. Dr. Kelley identifies most allergies with increasing parasympathetic dominance for several reasons: first of all, in these patients, the immune response in general tends to be exaggerated, leading to excessive reactivity even with mild exposures to offending substances. Then, Dr. Kelley proposes the membranes of parasympathetic dominant cells, including those of the immune system such as mast cells and basophils tend to be very leaky. The porous membranes easily allow irritating antigens to penetrate, and the mediators of inflammation such as bradykinin, histamine and serotonin to exit, provoking allergic symptoms.

"As the parasympathetic drift continues, this group becomes susceptible to angina, osteoarthritis, asthma, (a result of the bronchonstrictive action of acetylcholine, the main parasympathetic neurotransmitter), and skin disorders such as psoriasis. As they move still farther to the right of the horizontal axis, parasympathetics can fall victim to melancholic suicidal depression, severe cardiac arrhythmias and massive heart attack.

"The balanced metabolizers can experience problems and syndromes associated with either of the other two types, depending again, on their dietary choices. Someone with this genetic inheritance following a meat-based diet can end up with all the symptoms and syndromes brought on by an overly active sympathetic system; on a strict vegetarian diet he or she can experience the various problems associated with parasympathetic dominance".

MC's comprehension of this was to note that these autonomic nervous system categories could be altered by extreme vegetarian or carnivore diets. The efficiency could be altered by the consumption of more or less refined and depleted food. Without this comprehension and the ability to manage this crucial metabolic functioning, MC felt it would be very difficult to heal or know thyself.

It would seem that in ancient times, these categories were grouped and isolated from one another. This probably developed due to environmental selection pressures. Today, these groups have mixed with each other, which was the case with MC's immediate family.

TOULOUSE

To understand MC's brother Toulouse, one needs to know the effects of discouragement. Toulouse was one of the early war babies, a Baby Boomer from World War II. Festus, his father, had managed to survive the Normandy invasion as well as being on the front for a few months. Festus then received a small piece of shrapnel in the upper back thigh, or "in the ass," as Festus would put it. During his recuperation, he took an interest in Lucette. Just about the time the war was ending, Toulouse was on his way, making Lucette a war bride.

Toulouse's birth was difficult. The umbilical cord had wrapped around his neck several times. The doctor used forceps, which left Toulouse's bluish head somewhat misshapen for an hour or so. Post-war Europe was in shambles and Lucette always wondered whether the birth might have had some effect on Toulouse. Could this start on life have given him his reclusive and distrusting nature?

Sometimes described as aloof, Toulouse was a pensive and contemplative child who basically sat back and observed. His memories of childhood experiences were incredible. While MC had brushed aside most childhood memories, Toulouse remembered favoritisms and abuses in detail.

His lack of participation frustrated teachers and they would continually coax him to be a part of things, a trend that would follow him into adulthood. He developed quite an intellect and had a normal life style until his college days when a distracted environment weeded him out of college. The advent of the sixties led him to flee society and the draft and to become part of what he called the drop out generation.

Among Toulouse's traits was the onset of ruminating. This would take up a great deal of his time. During these times, he would begin to squint, chant and bob. This behavior, (almost a nervous tic), followed him all his life. Toulouse referred to it as a primitive ritual dance. He interpreted these movements as the manifestations of the betrayals in his life. It was a behavior he devised as a way of working through problems. As he got older, he was sometimes caught enacting this ritual and would suddenly compose himself. Most people realized that he was engrossed in private contemplations, something most are able to keep contained without grimacing and chanting.

One of the ways that Toulouse would work out this ruminating was to go for miles-long walks, sometimes throughout an entire night. He would balance himself through these marathons; sweat it out. This was among the times when his sympathetic nerves would kick into place. His balanced metabolism changed from parasympathetic dominant to sympathetic dominant during these metabolic swings.

Though he ate a fairly balanced diet, Toulouse often had cravings for different foods because of these marked metabolic changes. He would go from a week of eating only grains and wheat sprouts to eating sour dough bread to eating meat. These diet fluctuations drove him to the extreme sympathetic dominant state.

While MC and their father Festus tended to graze or pick and nibble, which often resulted in overeating, Toulouse ate regular meals.

What was most amazing at times was the quantity of food he could eat at one sitting. Afterwards, he would go into siesta mode, much like a lion after a large meal. Toulouse remained small and trim and some folks wondered where he put it all.

When visiting Toulouse in France, MC was told by mutual acquaintances about a sausage-eating contest that Toulouse had almost won. A dozen people, most of them large and obese, sat on a fairground stage eating sausages. Any sign of regurgitation or abruptly leaving the contest with a hand over one's mouth resulted in disqualification. At the end, there were only two contenders remaining: a very large woman and Toulouse, a slight guy with a ponytail.

The woman was able to finish the last offering by shoving it down with her fingers and was declared the winner. Toulouse had not quite finished his last sausage in the allotted time. He got up, collected his second place winnings, along with some leftover sausages, and went home to sleep it off. Afterwards, most of the other larger contestants vomited in the portable toilets but Toulouse had a strong digestive system and could have kept going. With more time, he probably would have won.

MC would later recognize that Toulouse was a balanced metabolizer. Thus, moods or nerves seemed to be beyond his conscious or voluntary control. Because of this, his digestion, respiration, blood circulation and secretions seemed to swing like a pendulum, bringing Toulouse somewhere in between a parasympathetic dominant type to a sympathetic dominant. This created a lifestyle that went from lethargic to energetic -- from sleep-in days to long energetic, walk-a-thon days.

Living as cheaply as possible affected Toulouse's diet. Since the cost of meat was hard on his finances, whenever he could get a free buffet of meat he ate his fill, perhaps too much at a time. For either economic reasons, or just plain gluttony, Toulouse and MC were both imbalanced toward the sympathetic dominant. Toulouse ate simply and was able to eat a large variety of things. He never rejected meat, which continually surprised people. Was it assumed that a guy with a ponytail who came up through the sixties would avoid meat?

MC was a sympathetic dominant and needed to balance himself with an alkaline diet consisting of vegetables, fruits and nuts. Without ever having studied these things, MC and Toulouse had a vague instinct about their respective, ideal diet types though neither of them followed their instincts very well. Dr. Mandel brought this concept to MC's attention during the cancer treatment. The doctor was a promoter of the statement, "One diet does not fit all."

In the case of Toulouse, Dr. Mandel emphasized that a balanced metabolizer who chooses not to eat a variety of plant and animal foods but adapts an extreme diet, either largely plant-based or largely meat-based, can push him or herself into either extreme parasympathetic or sympathetic dominant.

Although MC and Toulouse craved different types of foods, they were both eating a diet too acidic for their respective types. Dr. Mandel would later affirm that most cancer occurs in those with an out-of- balance diet. The healthy autonomic nervous system calls for what it wants. Finding a balance and acquiring habits is left up to the conscious thinking brain. As MC learned more about this, he would often witness people trying to impose diets on themselves that did not fit their autonomic nervous system. The guilt of slaughtering and eating animals is often behind these diet modifications.

Humans are divided into two branches: the sympathetic nervous system (acid type) and the parasympathetic nervous system (alkaline type.) As fate would have it, MC's father, Festus, and his mother, Lucette, were examples of these extremities. Lucette would be a perfect example of a parasympathetic dominant, or the alkaline type. When this PNS kicks in, the heart rate slows and the cardiac contraction weakens, while the arterioles of the skin and the digestive tract open widely. Blood pressure drops and blood flows more strongly into the skin and into the various digestive organs. The digestive activity along the entire gut picks up, including the secretion of acid, enzymes, bile and peristalsis. During this time, the parasympathetic nerves gear up and endocrine activity, such as the creation of adrenaline, is minimal.

Always on the go, MC and his father Festus were sympathetic nervous system-dominant or acid types. When the SNS fires up, the heart rate increases and the blood supply to the skin and the digestive system constrict while the vessels that feed into the muscles and the brains dilate. As a result, blood pressure rises and allows the blood to be preferentially shunted from the digestive system and the skin in order to favor the muscles and brain. During this fired up time, the sympathetic nerves inhibit the entire digestive process including all the digestive secretions as well as the muscular contractions of peristalsis that keep things going.
On the other hand, the endocrine organs are called on by the sympathetic nerves to release hormones.

This SNS was undeniably MC's type. Somehow, it was always easier for MC to see these traits in his father rather than in himself. Festus was often described as being "fired up." He barely slept at all and was constantly on the go, igniting himself with acid producing foods and creating a more acidic state. This could be practical if one has a task to accomplish. Otherwise, attentions might be diverted and the power of concentration diminished.

In MC's case, cancer seemed to love this acid state, especially in his colon.

Toulouse philosophically rejected the fired up state. He hated it in others and he hated it in himself. He strived for a more balanced and mellow environment and often pondered over it. "Why are we always in a rush? Why can't we stay in one place a little longer?"

He told MC that the spider was to be admired, "It sets up its web and then waits for its prey to come."

Toulouse's plant cultivations were based on a non-interventionist philosophy, which had some merit but was also problematic. His vegetation tended to choke itself out of existence. This way of seeing life gave the outside world the impression that Toulouse lacked initiative. He would have liked the outcome of his life to have as little to do with his participation as possible. His study of ancient sites, astrology, and the beneficence of wild plants seemed to fit in well with his balanced metabolizer mindset.

It was thought that Toulouse purposely behaved in a way that would counteract Festus's behavior. Toulouse noted correctly that his participation in projects with Festus would usually end up amounting to nothing and were sometimes a total loss due to Festus's inability to stay on course.

Festus would end up semi-dormant in a wheelchair, suffering from digestive and circulatory problems. He would be dumped by his second wife, stripped of all his things, and leave behind many unfinished projects.

Interestingly, Festus's prostate cancer would not be his downfall. It would be congestive heart failure, digestive problems, and the state of hypomania that would be his demise. All these were symptoms of his dominant sympathetic nervous system.

Toulouse seemed to have tendencies from both branches of the A.N.S. He seemed young for his age. He seemed to have no eating disorder, except for participating in the occasional sausage-eating contest.

Unlike his father, Toulouse rarely ever started a project. He kept the area that he utilized clean. Other areas of the house that he did not use were kept exactly as they were.

He would collect some things, mostly jars and containers for storing food. His admiration for spiders was quite apparent, as he never interfered with their web making, no matter where they placed themselves. Toulouse occasionally would eat cooked grains for days at a time. A parasympathetic could not have fared well on this.

MC was aware before his colon cancer that he was not eating properly. This was before he had any inkling that people might need different diets.

One spring, while visiting Toulouse in France, MC was introduced to an example of a very alkaline diet. Toulouse had knowledge of edible plants. He was preparing a salad made from stinging nettles, dandelions, sprouted seeds and wild shoots. MC could instantly feel and recognize a strengthening effect from eating the salad.

In the Kelley model, the sympathetic dominant pancreas, of the three types, is the most inefficient, synthesizing the least amount of the various digestive enzymes due to the suppressive action of the sympathetic nerves on the organ. However, a predominantly vegetarian diet will inhibit the overly strong sympathetic division and stimulate the weak parasympathetic nerves, helping, in turn, to activate the pancreas into greater productive efficiency. Since plant foods can be digested quite effectively with even minimal quantities of most enzymes, as long as a sympathetic dominant follows the appropriate diet, Dr. Kelley claims the pancreas should produce sufficient proteases for both digestion and protection against cancer.

However, MC was taken over at the time by the availability and variety of foods in France. This was the wine and cheese culture. Add to this some crusty breads and chocolaty confections and MC was using most of his energy trying to digest this acid-producing mélange. Bound to these foods by cultural gluttony and childhood attachment, MC would feel stuffed every night.

Twenty years before, MC and Toulouse had imposed sanctions on their diets in the form of fasting. They would both undergo the ten-day water fast.

(Paul Bragg, "The Miracle of Fasting")

Fasting triggers the autonomic nervous system to adjust. With nothing to digest, the body is allowed to go through drastic elimination. Most people think they could never do this, although animals seem to instinctively fast when they get ill.

For MC, it was easy, cheap and natural. It was draining at times but it could also be tremendously energizing. His thought process calmed, his concentration intensified and his blood pressure and heart rates normalized. The disappointing part was falling back into one's old eating habits when ending the fast. At the time, both MC and Toulouse were unclear about individual balanced diets. They would inevitably go back to their pre-fast way of eating. MC and Toulouse never exceeded the ten-day water fast. For MC, it was initially a way to clean out and lose a lot of weight.

MC remembers reading about the fasting clinics that were used to treat cancer. When a fast exceeded forty days, the metabolic system began to consume body tissues. Vital organs such as the brain, heart and lungs were spared this initial induced starvation period. Any remaining minerals, salts, fluids and some muscle cells were then metabolized for nutrients. It was during this phase of the fast that tumor cell tissues, growths and unessential foreign proteins were cannibalized, as well. The results were dramatic and successful. However, after this great effort, there were some eventual reoccurrences of tumors. Simply starving cancer away was not dealing with the imbalances and causes. MC and Toulouse had fasted when they were in their thirties. It dropped from their routines and was not utilized again until MC began the cancer treatment.

On the cancer treatment, MC was now undergoing twenty days on his prescribed diet and supplements. This was followed by a five day break before starting the next 20 days of his moderate vegetarian metabolizer diet.

During the five day break, MC would do, in sequence, Dr. Mandel's prescribed protocol:
- Liver flush protocol
- Clean sweep protocol
- Citrus purge
- Carrot juice fast

This prescribed routine put the body at rest and aided in the rapid removal of metabolic waste from the body and the rebuilding of damaged tissues. MC found it interesting that the cancer treatment's 5-day break put him in a similar state of mind of that when he fasted.

The citrus purge involved two days of drinking doses of Epsom salts and water followed by drinking a gallon mixture of grapefruit, oranges, lemons and water throughout each day. Yet MC felt the same mental vigor that he had experienced after undergoing a ten-day water fast.

The liver flush protocol was a simulation of the older 3-day apple juice fast, yet MC ate his normal prescribed alkaline diet. A gallon of apple juice was mixed with one ounce of phosfood liquid to be drunk in four days, 1 quart a day.

After ending the protocol on the fifth day with, among other things, Epsom salt, berries, whipped cream and olive oil, MC felt dynamic. For him, this particular protocol was one of the turning points of the cancer treatment. He experienced the same deep sleep he had experienced during the long water fast. Also, while eating his prescribed diet, his body discarded the remnants of his gallstones.

The three day carrot juice fast, (that contained mostly carrots, some celery, apples and a little ginger), involved drinking 8 cups of this mix on each of the three days. MC felt the removal of waste through his skin.

When done in coordination with the twice daily coffee enemas, (morning and night), MC was promoting detoxification.
The coffee enema consists of one quart, (2 tablespoons), of ground coffee to one quart of boiling water. This is divided in half, cooled and then held for 10 minutes for each pint.
Once again, it was quicker and more efficient than the skin eruptions that sometime occurred during those longer unguided water fasts.

In more recent times, through his 12th year of the cancer treatment, it is sometimes possible for MC to extend the mini fasts during these five-day breaks. It tends to accentuate the effect of MC's alkaline diet for a sympathetic autonomic dominant. It was far easier than the extended water fast, which would sometimes result in skin eruptions and often be followed by impulsive cravings, compulsive eating and gluttony.
Even though fasting never really changed MC's life or spared him from cancer, it did give him the experience that

he could use the effects of his metabolism to alter his outlook and wellness for at least a short time.

Because MC had experienced such fasts, the cancer treatment protocol seemed mild by comparison. MC could now simulate the same beneficial effects of a long water fast without the extreme deprivation. Inadvertently, it was preparation for things to come and it confirmed the statement, "We are what we eat."

It was during this intentional hunger that MC would crave the things he needed. MC imagined that his body was more receptive to its prescribed supplements and diet after such purges and fasts.

By fasting, was MC asking the autonomic nervous system to regulate itself? By depriving his body of an inherent craving, was MC forcing his somatic life cycle to find its own equilibrium?

During their fasting days, MC and Toulouse were testing the limits of their bodies hoping that it might help them change their outlooks on life. Fasting was not a means to an end. The hunger of the body engages the autonomic nerves for better or worse. It ultimately showed MC that these autonomic nerves could be appeased or stimulated by different foods or by the lack of food altogether.

During pre-cancer days, MC would frequently run solely on adrenaline and eat a steady stream of foods that did not necessarily complement each other. On one occasion, he ate over a pound of chocolate along with pickles and other convenience foods. Even though MC did not eat red meat or drink whiskey, he would eventually be diagnosed with gout. He was later told that this makeshift uric acid-producing diet had made a perfect setting for his colon cancer.

Toulouse, by contrast, enjoyed a wider range of more acceptable foods. He could eat plain cooked grains for days. Parasympathetics do not do well on such a diet. On the other hand, he could also eat foods high in acid along with meat and peanuts, foods that do not fare well with sympathetic types. Overall, Toulouse was noted for eating fairly wholesome meals.

However, it could be said that neither of the two brothers was eating in concert with their true metabolic types. They were both eating too acidic a diet.

Toulouse had come to the U.S. to help MC care for their father, Festus. This would last a couple of years. Part way through, Toulouse developed what he described to his friend Madonna as a persistent flu.

It had been a stressful two years with their ailing father. After Festus's death, Toulouse revealed that there was an effect or sensation in his eyes that was not going away and that his vision seemed altered. He thought he might have glaucoma. Toulouse realized that he needed medical attention but found himself bogged down with paperwork. He had his medical card from France but he was trapped in a land that had no universal health care. Soon, Toulouse was asking people what to do about migraine headaches, which he had never suffered from before.

Toulouse had planned to house sit for MC. A few weeks had gone by without much contact between the brothers. MC had tried to call Toulouse but there was never any answer. Soon, Toulouse's friend Madonna called MC to ask if he knew why Toulouse wasn't answering his phone. MC was leaving for the weekend and was desperately trying to get hold of Toulouse. Finally, after endless ringing, Toulouse picked up the phone. He was disoriented and said that he had heard the phone but was unable to get to it in time. MC reminded Toulouse that he had promised to housesit and that he was supposed to be there already. Toulouse replied in an unusual way, alerting MC that something was amiss. Toulouse told MC he would immediately leave for his house.

Toulouse had a horrendous trip driving himself to MC's. When he had almost reached the house, he took a wrong turn and was lost for a few hours. MC kept calling his house to see if Toulouse had arrived. Finally, MC called a friend to look into the situation. The friend confirmed that Toulouse had arrived but, again, he did not seem able to get to the phone. The friend mentioned that when she found Toulouse in bed, he appeared dead. When he snapped out of it, he communicated somewhat normally.

When MC returned home, he saw that none of the chores Toulouse had agreed to do had been done. He found Toulouse in bed. He had managed to knock down all the lamps in the room. The covers and sheets were tangled around him. As soon as Toulouse tried to get up, he appeared to be intoxicated.

At first glance, an alternative practitioner saw something wrong toward Toulouse's forehead. At one point, he developed the hiccups, which made his condition seem even graver. Occasionally, his eyes went into a stare as if he was having a mini-seizure.

During a warm spell just before Thanksgiving, (the fifth anniversary of MC's colon cancer removal), Toulouse was rushed to Hilltop Hospital.

Something was located between the lobes of Toulouse's brain that was causing pressure and plugging ducts that would normally allow fluids to flow around the brain. It was located in a bad spot. Toulouse was admitted to the hospital.

MC called Dr. Heil Mittel Mandel and, to MC's surprise, Dr. Mandel was especially interested and concerned.

Up to this point, MC didn't know about the different types of cancers and their corresponding autonomic nervous systems. Toulouse's predicament would make MC aware of these details. From the description that MC had given him, Dr. Mandel was identifying Toulouse's type of cancer.

The doctor said, "Toulouse could respond very well to the enzyme therapy. But he will need to be stabilized so he can do the protocol."
Dr. Mandel seemed to know this intuitively without ever having seen or met Toulouse. MC soon learned the reason why. By knowing the type of cancer someone has, his or her autonomic nervous system imbalance can be determined. The two branches of this system, sympathetic and parasympathetic, tend to manifest different types of cancer.

MC and Toulouse's mother, Lucette, who died of brain cancer, was a parasympathetic dominant. She escaped most digestive problems but had a tendency to allergies, asthma, chronic bronchitis and chronic fatigue. This type is prone to immunological malignancies such as leukemia, lymphoma, Myeloma and in her case, Glioma of the vermis in the left cerebellum hemisphere, which controls parasympathetic functioning.

MC, the sympathetic dominant, would be prone to the digestive diseases such as irritable bowel syndrome and anxiety but less prone to depression or allergies. Sympathetic dominants are prone to the common solid tumor such as colon cancer or cancer of the breasts, lungs, pancreas, liver, uterus, ovaries or prostate. Toulouse had traits from both branches. Even though MC had been following a protocol to try to balance out his sympathetic dominant tendencies, it was something he was now only beginning to grasp.

Meanwhile, in the hospital with a cloud of urgency about him, Toulouse was stabilized. Hilltop was ready to intervene with major brain surgery. While doing more x-rays they found a mass wrapped around the outside of Toulouse's kidney. It was this mass that had spread to the inner depth of his brain, not inside the brain but between the lobes of the brain. The surgical plans were cancelled as soon as the cancer was discovered to be on the kidney. As it turned out, the cancer had started in the kidney; a situation known as renal cell carcinoma to the brain.

Toulouse responded well to MC's advice, especially with the backing of Dr. Mandel. It was obvious to MC that the only hope for Toulouse was to internally dismantle or shrink this growth. At that point, MC was not as knowledgeable about the effectiveness of the pancreatic enzymes. Despite this, he brought a small bag of these enzymes to the hospital for Toulouse.

A nurse holding a zip lock bag containing the enzymes addressed MC the next day. She explained that nothing was to be given to Toulouse, "No vitamins!"

Toulouse was given steroids, which made him seem fairly normal. Stabilized a bit, he was able to eat. As had often been said in the past, one of the nurses replied, "Toulouse can really eat!"

MC had been made power of attorney in a flash. He told Toulouse that as soon as he was more stabilized, he could go home and start the cancer treatment with Dr. Mandel.

Toulouse was pleased with MC's statement.

He said in a pensive way, "I guess I'm taking over where Dad left off." Their father had died the month before.

MC told Toulouse that there needed to be a different way of dealing with this situation. In order to get home and begin Dr. Mandel's treatment, Toulouse needed to abandon his old thought system and reorient his faith. In so doing, he would instruct his body to stabilize this "head thing"

MC explained to Toulouse that there was a way to change one's mind and put one's trust in these new decisions. These words were spoken in a direct and sober manner. Toulouse agreed with MC by slowly nodding and then was quiet for a while.

In an inquisitive and sincere way, he asked, "Have you ever been able to do this?"

MC's mind went still. There were a few seconds of silence between the two brothers and then MC heard himself say, "Yeah, once." MC realized that this impromptu advice had been his personal experience.

Toulouse was far worse off than MC had been prior to his colon surgery. He had waited a long time with this ailment. This procrastination was a character trait that the brothers shared. Now it was urgent that Toulouse make a leap of faith.

The discovery of Toulouse's kidney cancer had stopped everything. The hospital was truly at odds about what to do. There was a half-hearted proposal to try a procedure using radiation called "Stereo Tactic Radial Surgery" where a series of weaker beams that crisscross at the tumors' locations might destroy the tumors without destroying the surrounding brain tissue.

Throughout his life, Toulouse had always been discouraged by the contests in society. He disliked and avoided them. He was now totally dependent on the decisions of others. He was trapped. He had lost his opportunity to make a decision to change his destiny.

Discouragement, criticism, disappointment, failure and shame, all things Toulouse had successfully avoided dealing with, was the perfect brew for creating kidney problems.

Before MC left that evening, he promised Toulouse that as soon as this treatment was over he would take him back to France.

Toulouse had rekindled an old relationship with Madonna. He shared his difficulties with her, as he became more and more preoccupied with his head problems. He got great comfort from talking to Madonna each evening by telephone.

Just before MC was leaving, Toulouse got a phone call from Madonna. Toulouse was well occupied by the phone call as he waved goodbye to MC. Madonna had been an intimate acquaintance to Toulouse maybe twenty or thirty years prior. She had told Toulouse at that time that she had kids and a husband to deal with. Still, Toulouse kept a correspondence throughout the years.

While taking care of his father Festus, Toulouse learned that Madonna's kids and husband were gone and her parents had passed on. The two had many common interests and Toulouse was able to take a trip to visit Madonna around the time his father had died. Before Toulouse collapsed at Hilltop Hospital, there was even a plan for them to get back together. Madonna gave a great deal of comfort to Toulouse.

MC would find a letter from her written to Toulouse during the separated years.

Toulouse – There is something I must tell you. For a long time I have prayed to God to give me strength to end my life on earth or send me someone who could give me love and take away the aloneness. I believe he sent you because you understand what loneliness is. I am sure you have felt this, even though you cannot believe it as yet. Do not be afraid; I will not hurt you. Neither of us will have to be alone again. We can give each other a love we both desperately need, whether we are together or alone.

Peace be with you,

Madonna

MC wondered how this sort of trouble could happen to Toulouse during such a romantic reunification. That night at the hospital, Madonna's phone call lasted a long time.

Toulouse finally told Madonna, "I want you to come to France, but I want to go ahead of you. MC's going to take me there soon."

Madonna said, "I've never gone so far by myself in a foreign country."

Toulouse then explained to her how easy it was. He described the trains to take, the right turn at the station and the climb to the house. He gave her explicit instructions. He described the view of the wheat fields and the wonderful smells when it rains.

He told Madonna, "When you get to the house, just go in. I'll probably be taking a walk in the fields. Make a fire in the stove and just lie down and rest."

Madonna later told MC that some of these directions were as if he were there. Most of the conversation was normal, but the part about his leaving ahead seemed a bit odd to her.

The next morning, MC received a call from Hilltop Hospital.

"Everything is okay," he was told. "We just want to do a kidney biopsy on Toulouse."

MC now had the power to decide such things and it made him feel very uncomfortable. He said, "Well, I thought there was no question as to what it was. Dr. Ming told me she knew what kind of cancer it was. I am fine with not doing it. Don't feel you need to do this for our sake."

The surgeon replied, "We need to know what we're dealing with. Since open surgery is being replaced by Stereo Tactic Radial Surgery, we won't have the option of taking a biopsy sample. We need to get a sample from his kidney instead."

Nothing had been done to Toulouse for over a week and the hospital was getting restless.

MC felt trapped. "I won't be able to get up there very quickly to sign papers," he said.

"There is no need; your permission over the phone will suffice. There's no rush to get here. He'll be out and coming to at about eleven this morning. There are rarely any complications with this procedure," said the surgeon.

MC arrived at Hilltop Hospital just before eleven and made his way to the unit where Toulouse was located. It was an open, octagonal space with monitoring devices in the center of the unit. As MC spotted Toulouse, he noticed the nurses hovering over the bed. Toulouse was squinting as he rubbed his forehead with his right wrist in a repetitive, pawing gesture. One of the nurses was trying to discourage him from doing this. MC's approach seemed to make the nurses a little self-conscious. One of the nurses drew the curtain shut and asked MC to give them a minute. MC gently moved alongside the edge of the room and slid around the curtain until coming up to the foot of Toulouse's bed. He became pensive and still which made his presence less noticeable, even though he was inside the curtain.

MC began to realize that Toulouse had been struggling like this for a while. The hand and wrist that he was using to rub his forehead was wrapped with protective gauze. The nurse informed MC that Toulouse's blood pressure had begun climbing since coming out of the biopsy procedure. "It's causing some internal bleeding," she said.

MC was riveted at the foot of the bed as Toulouse continued to frown and paw at his forehead.

Then the nurse started talking to Toulouse, close up and as though he was far away. "Toulouse? Mister Mac, can you tell me if you're...?"

Toulouse opened his eyes, but had the stare of someone not seeing. His legs went stiff with his feet pointing straight down while his head arched back.

The nurse said, "He's having a seizure."

She again suggested that MC step outside the curtain. MC was frozen in his spot. He was hardly aware of his body or his emotions. All cordialities seemed pointless and absurd to him at that point.

There was an attempt to drill into Toulouse's skull to alleviate any possible build up of pressure that may have occurred during the biopsy and the swings in blood pressure.

MC could not recollect if he had given the hospital the okay to do this At this point, it would have been just a formality. As far as MC was concerned, Toulouse was done.

While they were piercing Toulouse's skull, MC drifted through the nurse's station and saw a scan of Toulouse's brain on all the computer screens. It showed the central core of the brain to be compacted and dark.

It was explained to MC that the brain exudes moisture on a continual basis. The fluid needs to drain through ducts that are located near Toulouse's metastatic tumor. The constriction along with the build-up in pressure is what caused the Grand Mal seizure.

After the drilling procedure, MC sat on a chair near Toulouse's bed. He was convinced that the biopsy had triggered the seizure. MC had given permission for the biopsy without even having a final communication with Toulouse. He knew Toulouse's temperament. MC imagined the apprehension Toulouse might have felt towards the biopsy which could have triggered his sympathetic side and ignited a fight or flight reaction. The vessels that feed into the muscles and into the brain must have dilated. The stronger flow of blood would raise his blood pressure and increase blood flow to the brain. The surgeons saw no correlation between the high blood pressure and the biopsy procedure. Nonetheless, these were the imaginings occupying MC's thoughts.

"This is exactly what you needed to avoid," whispered MC to Toulouse.
There was still a slight pulse in Toulouse's body. MC felt tremendous guilt as his thoughts were preoccupied with Toulouse's post-mortem affairs.

MC's stare turned toward the window as his sadness overcame him. It was as if Toulouse had waited for MC to get there before having the seizure. Now Toulouse rested peacefully, lying very still with a white turban-like bandage around his head. Soon, he was placed in the intensive care unit. MC was told to wait until things were settled before sitting with his brother.

Surgeon Ming saw MC sitting in the hall and gently approached him.

She was dressed in her light green scrubs and with a beautiful compassionate smile, said, "I guess Toulouse took care of this one himself?" She went on to explain that any intervention would have caused severe damage to Toulouse.

The day after the seizure, MC would have the life support apparatus on Toulouse removed. This was the final member of MC's immediate family to die, all after having some bout with cancer.

MC started drifting back to his mother's death. Toulouse and MC's mother, Lucette, had died some thirty years before. In her case, an operation had precipitated her death. Lucette's illness started a year before her final surgery. Perplexed by her case, the doctors' diagnoses jumped back and forth between a stomach ulcer and a gall bladder problem. There was eventually surgery on the stomach and the gall bladder. Though nothing much was found in either spot, the gall bladder was removed anyway.

After barely a few months, Lucette showed a swelling at the temples. A tumor in the hindbrain was discovered. It was described as not being a solid tumor (parasympathetic dominant) but more like a penetrating blotch, as if a drop of ink had soaked into a snowball. Dr. Mandel would describe this as an immunological malignancy.

MC had gathered his grandmother to come and see her daughter, Lucette, just before the surgery. A few things stuck in MC's mind. One was that Lucette improved drastically a day or so before the surgery and seemed normal as she was brought to surgery.

It was announced at the termination of the surgery that Lucette would not survive. MC's grandmother was frantic. MC thought it might help if he took her to visit her daughter in the intensive care unit. It did help his grandmother. She had imagined devastating effects and was relieved to see Lucette looking quite whole with a small, bandaged turban around her head.

MC's grandmother was relieved by the visit and said, "At least they did not butcher her."

As MC and his grandmother walked in, Lucette was actually looking around. Her limbs were moving rather slowly and in an uncoordinated manner. When her gaze fell on them, the monitor waves started to change drastically. Lucette then slowly lifted her right arm and brought the back of her hand up to her face as if to hide from them.
MC's perception was confirmed when his grandmother said, "She sees us. I think we're bothering her. Maybe we should leave?"

Lucette put down her arm and looked off in another direction. Her eyes looked dilated yet pensive.
MC could see that the monitors were still a bit turbulent. As they left the unit, MC told one of the intensive care nurses that he thought the visit might have agitated Lucette.

The nurse told MC, "I don't think so. She has been unresponsive to any of my questions or tests."

A few hours later Lucette would be gone - lying still with all the monitors off.

The surgeon would then speak to MC and his father, Festus. MC noticed right away that the surgeon could not make eye contact with him and either looked away or looked at Festus. Festus would not make eye contact with the surgeon.

The surgeon said, "We were hoping to arrest the tumor so that we could do some radiation in hopes of prolonging her life for maybe six months. She would have been totally paralyzed from the neck down because of the surgical removal of part of the vermis."

MC was not involved in the decision to do surgery and wondered why this information might not have been shared before the surgery.

At a later time, MC would always wonder if his mother's case was an acceptable form of euthanasia done for humanitarian reasons. It was probably what gave MC and Toulouse an underlying distrust in medical procedures. At least Dr. Ming professed there was nothing that could have been done surgically for Toulouse.

7

LUCETTE

MC's mother Lucette had a "parasympathetic autonomous-dominant" nervous system.
She always complained that her hyperactive husband, Festus, (a sympathetic-dominant), was an "insane man."

Lucette was not terribly happy and was often depressed. She had immigrated to the United States as a war bride. She often associated the ways of this country with her husband's hyperactivity. MC wondered if knowledge of the sympathetic and parasympathetic dominant nervous systems might have helped Lucette in harmonizing her relationship as well as avoiding cancer.

If MC had been limited to his own reflection, he would not have been sure of his mother's A.N.S. type but, as Dr. Mandel would confirm, she got a cancer that can only be acquired by a parasympathetic dominant. Her final diagnosis was Glioma of the Vermis. According to Dr. Mandel, this was one of those immunological malignancies that can only inflict those who are parasympathetic dominant. This cancer's location, in the left-cerebellum hemisphere, (the hind brain), is responsible for parasympathetic functioning. In the thirty years between Lucette and Toulouse's deaths, not much had changed in the medical world's approach to treatments or explanations.

From "One Man Alone: An Investigation of Nutrition, Cancer, and William Donald Kelley", by Nicholas J. Gonzalez, M.D.

"According to Dr. Kelley, in parasympathetic dominants, the pancreas tends to be durable and quite efficient, capable of manufacturing copious amounts of all the various digestive enzymes. Consequently, this metabolic type tends to be protected against the classic hard sympathetic malignancies, such as breast, colon, and lung cancer, which don't stand a chance, surrounded as they take form by large quantities of the pancreas proteases. However, a parasympathetic following a plant-based diet can develop certain immune-related malignancies despite the usually efficient pancreas. Since the parasympathetic nerves stimulate our various immune cells and organs such as the thymus, eventually, this system can end up in a highly overactive state. Should the extreme parasympathetic drive persist, it will override the protective influence of the circulating enzymes, leading to cancers of the immune cells such as Hodgkin's disease, leukemia, lymphoma, [Glioma of the Vermis] and multiple myeloma."

MC was often reminded that Lucette and Toulouse both had brain tumors, which must suggest a genetic link. In a simplistic way, this sounds true. However, other than the fact that Lucette and Toulouse both manifested a form of cancer inside the skull, the link was inconclusive. MC knew enough about them, their temperamental differences and their two different cancers, to be unconvinced of a genetic explanation. A genetic explanation for cancer seemed to MC like a smoke screen to give people a theory that would release them from any responsibility. MC was somewhat resistant to the news always linking human's problems to genetics.

Within this same period MC would often hear news blurbs proclaiming smoking, stuttering, A.D.H.D., eye degeneration, obesity, longevity and, of course, cancer, as examples that could be blamed on the human genome. It was even suggested to MC that there was a gene for stinky feet.

This seemed like a simplistic way of alleviating guilt from any involvement having to do with these ailments, behaviors or deficiencies. To MC, it interfered with the possibility of discovering real explanations. This genome mapping seemed to have come about not only to wash one's hands of any self-responsibility, but also to prop up a faltering germ theory that cannot explain the spontaneous occurrence of degenerative diseases and the bugs that accompany them.

Cancer science seems to be drifting further into the intricacies of this programmed genetic make-up. Thus, there are postulates to create blocking agents, which often have side effects. These agents are aimed at specific genes that are supposed to be the cause of particular cancers. The lack of breakthroughs is often blamed on mutations in the genes, which will then require more research. While the industry is busy creating more patented drugs in an attempt for patients to better tolerate their already patented and ineffective toxic drugs and chemicals, MC began to see the endless complexities of scientific research.

It was after Toulouse's death that, other than blaming genetics, a more comprehensible explanation for the origins of cancer would make its way to MC.

MC's results from enzyme therapy seemed unbelievable to those who have been force fed forlorn cancer statistics for ages. He had been given a year to live. Ten years later, his surgeon, Dr. Navaja, referred to MC as "miracle man." MC was becoming aware of the accelerated pace at which he was burying family and friends from a disease that was considered incurable and yet, in an attempt to retard its spreading while offering palliative care, considered treatable with chemicals and radioactivity. One of the definitions of palliate is "to cover by excuses and apologies."

It was a shock for MC to learn that a newly rediscovered explanation for cancer, the trophoblasts as its origin, was over a hundred years old.

John Beard was an embryologist from Scotland who, in the late eighteen hundreds, began his interest in cancer. His investigation started from the reproductive birth process. Beard was studying asexual cells in plants and concluded that all creatures must have a form of these, however small or for however short a time. He finally identified the surrounding cells of the embryo to have these similar properties. They seemed to guide the development of the embryo. Beard referred to these trophoblasts as, not only being able to invade into the uterus wall, but of directing embryonic growth itself. Along with its invasive quality, it was also responsible for the creation of the placenta. In John Beard's time, the placenta was considered to be a sample for cancer research. If the fetus died at a particular stage, the mother could be invaded by the proliferating placenta and could die from the cancer called choriocarcinoma. Beard concluded that the appearance of the pancreas in the fetus controlled the growth of the placenta. Beard then affirmed that cancer growth could be slowed by being injected with the pancreatic enzyme. Great efforts were made back then, as are made now, to prevent or censor any possibility that this simple explanation receive any recognition.

Dr. H.M. Mandel resurfaced with John Beard's findings and, along with Dr. William Donald Kelley's studies of the autonomic nervous system, began treating cancer. MC's life was spared because of their work. Yet, when MC described this to people, they were incredulous. How could something this important have remained so obscure?

However, John Beard pushed onward. Instead of trying to see any fault in the DNA, he noticed that the activity of the DNA was regulated by the trophoblasts. Without the trophoblasts, human reproduction would cease. This is still one of the last great mysteries in reproductive biology.

Beard went further. He described how certain cells migrate inside the soon-to-be fetus and lodge in random places on their way to the genital ridge. These dispersed cells, presumably stem cells within these trophoblasts left over from the earliest embryonic formation, are capable of invasion, migration and angiogenesis. They are scattered about in order to repair the body. Occasionally, one of these vagrant trophoblasts will be triggered to produce a fetus-like growth called a teratoma. This cancer's composition of skin, teeth, eyes and hair would tend to associate it to the early embryonic developments. Some superstitions describe the phenomenon as one's twin wanting to be born.

This trophoblast resolution gave MC incredible resolve that he had come across a science that had insight into cancer and its origins. The trophoblast of the human embryo produces the placenta, which is similar to cancer. The fetus then produces pancreatic enzymes to stop the growth of the placenta and then manipulates it for its own nourishment before discarding it at birth. If the trophoblast loses connection with the embryo, gestation comes to a halt. When a trophoblast, dispersed within the body, is called upon to repair a troubled organ, and for whatever reason produces cancer, a similar embryonic law can be applied.

MC's ingestion of pancreatic enzymes brought his body to a balanced state in order to appease and dismantle the cancer producing trophoblast. These were the essential steps in his recovery. MC felt that the slippery slope of status quo cancer treatments only succeeded in replicating more and more troubled trophoblasts, causing them to be released into the blood stream to create further carnage. MC was his own proof and needed no persuasion as to whether John Beard's findings were valid. Gone was the idea that a mature differentiated cell of an organ would just mutate and start uncontrollably dividing with distorted DNA. It made sense to MC that the distorted DNA was the result of cancer and not the cause. Gone was that void feeling upon hearing defeatist explanations of a predetermined genetic make-up. MC was convinced of how much these beliefs prevail when hearing of young women surgically removing their breasts to control breast cancer that "runs in the family". How then could MC have healed from a mutated disease with a programmed genetic code leading to certain death?

Why were there so many expressions of cancer, hundreds of different kinds?

How is one area selected over another?

Why was cancer, as Dr. Mandel told MC, reaching epidemic proportions?

Why did people temperamentally different in autonomic nervous type acquire different types of cancer?

Why was cancer increasing in countries whose diets have modernized?

Why were cancer rates directly correlated to the amount of exposure to electromagnetic and nuclear energy?

Now there was a rediscovered science and an encouraging voice that told MC he could overcome cancer and thus uncover the cancer scourge. The knowledge of the origins of cancer would lead him to the disallowance of cancer. MC needed to recognize and accept his sympathetic dominant tendencies and counter them with the cancer treatment.

From "One Man Alone: An Investigation of Nutrition, Cancer, and William Donald Kelley", by Nicholas J. Gonzalez, M.D. --

"In his model, Dr. Kelley describes the pathophysiology of cancer in particular detail. In general, for very specific reasons he associates the "hard tumors," the common malignancies of the breast, colon, liver, ovary, pancreas, or prostate with extreme sympathetic dominance.

"Most conventional cancer researchers propose that the disease develops from normal cells in any of our various tissues, the end result of defects in the DNA occurring spontaneously due to exposure to environmental toxins or as part of the normal aging process. At times, the thinking goes, these abnormalities or mutations can lead to malignant change. Dr. Kelley believes, however, that cancer forms only from stem cells, undifferentiated precursors located in all our various tissues, that serve as replacements for those cells lost due to normal turnover, disease, injury, or aging. Though they serve a most useful function, without appropriate control of their replicative tendencies, they can form invasive tumors."

If these vagrant germ cells within the trophoblast are lodged within the body the question is, why would these cells start invading, growing and migrating? Dr. Beard continues by noting that these germ cells, which today are called stem cells, within this irresponsible trophoblast are capable of imitating or mimicking the various organ tissues; thus explaining the medical world's widely accepted theory that the cancer cells must originate from the mature organ cells.

MC began to wonder if a self-destructive wish could induce a trophoblast to produce deadly invasive cells. As far as MC was concerned, his brother Toulouse and his mother Lucette led troubled existences. MC remembers his own irritable behavior along with the irritable bowels. MC's recollection was that the setting for his colon cancer probably had been established several years in advance, and that his perpetual ruminations about family betrayals and business torments had made a perfect setting for the conception of his cancer.

A general ingratitude for life's events and a preoccupation with troubles were traits Lucette might have passed on to MC and Toulouse. When MC was in his adolescence, Lucette once mentioned how courageous it must be to end one's life like Puccini's Madame Butterfly. MC mused about the enflamed cells of the body giving a signal to the trophoblast to do something in desperation, only to be misguided by the lack of a will to live. It was critical for MC to have had this drastic terminal situation to bring about his reflex for survival.

While observing MC with his vitamin pill bags, coffee enemas and special dieting requirements, a friend said to him, "You must really want to live to do all this work."

MC did not perceive this as work. It had initiated a new desire to live

Perhaps MC's autonomic nervous system type, the sympathetic dominant, was determined genetically, but his will to apply correction and to become aware of his body's needs was an act of self-determination.

MC was not simply interested in treating symptoms that would eventually take his life - he was interested in using his symptoms to discover his true nature. MC realized that these symptoms were not from his genetic code but from his body's voice pleading for a change of vision.

The greatest memory of MC's life to this point was the witnessing of his condemned body coming back to life. Toulouse missed his opportunity to do the cancer treatment partly because of his procrastination and partly because of the default in society's cancer science.

How many people will perish because of the cancer metropolis's inability or unwillingness to take into account the variances of people's autonomic nervous systems?

When will it be acknowledged that the autonomic imbalances and autonomic inefficiencies are the root cause of cancer?

When will it be recognized that the variance of vitamin and mineral complexes vary for each individual patient?

How long can the use of "like tissues" (photomorphogens) such as pancreatic enzymes as a healing tool be overlooked?

It has been over a hundred years since embryologist John Beard discovered that a similar mechanism to one's reproductive forces is at the origin of cancer. He would also discover that these same procreative dynamics could be guided to dismantle cancer.

8

SHARINIGAR: WHERE THERE IS TRUST, THERE IS FAITH

In his early forties, MC would encounter a man, Sharinagar, who would have a profound influence on him.

It was a perfect fit for MC, who had been disillusioned with dogma, rituals, organized religions and "isms." Sharinagar didn't go beyond MC's tolerance for esoteric spiritual interpretations. Such topics as these, voiced in a more pious setting, would always seem to MC like a bible study class. Sharinagar, seen as a great integrator, was often viewed in a negative light because of his knack to unravel ages-old, mainstream truths, leaving no stone unturned and no teachings unscathed. He intuitively brought up lessons that were pertinent to the individuals present at his retreat.

Sharinagar's prophetic ways drew people to him but his austerity and frankness drove away the skeptical hordes. "Why do you come to listen if you never make the decision to apply this to your life?"

MC was convinced that there could be a transferable energy between people but that most chose to live an isolated, self-absorbed existence. Sharinagar referred to this as 'separation'. To Sharinagar, this widespread separation was the root cause of all humankind's dilemmas. He had a way of melting the great void that MC felt. There was an experience of atonement during Sharinagar's sharing. It was like being uplifted from the insanity of the busy world, at least for a short time.

Sharinagar proclaimed that humankind was behind in its evolution and there was urgency to catch up. A celestial speedup was put in place in these times to intensify the awareness of the problem. What most of the world considered the "real world" was, according to him, an unreal and separated world, giving meaning to meaningless things.

MC began to realize that he was trapped between his real and unreal worlds and struggling to reconcile the two. There were occasional ominous statements made by Sharinagar that a truth once heard and recognized but not applied, would often turn into poison.

Sharinagar had never been formally educated. He once noted that he had less book knowledge than any of the people who came to listen to him. During middle age, he began to express an innate insight. His discerning vision was coaxed on by his affinity for the universal array of spiritual wisdom. He refused to own property. He never succumbed to religion, racism, nationalism, politics, scientific belief or fashion. Thus, he seemed to have a lot less attachment to his learnings or his cultural background. He had spent five years in silence, which helped sensitize him toward his self-ordained function: To help others find their silence.

However, he did not recommend that others follow in his steps. He had wandered in the Himalayas and though he had most likely delved deeper into himself than most, it was not a guarantee to being awakened. This, he concluded with some disillusionment.

He warned against quitting one's job or changing "the externals". He stressed that only internal change would bring about peace. This would happen without mind control, mantras or preaching.

Sharinagar felt that one should always give people space: "Leave other people alone".

For MC, Sharinagar's approach to teaching how to apply spiritualism to one's life was easy to understand. This was learned by first exposing and undoing an unreal, earth-bound existence that, by self-evidence, was exacerbating our inability to feel inner peace. The inner world of silence was always accessible and visiting it often was the key to one's liberation.

MC never got angry with Sharinagar or this higher form of power. He realized that it was his own doing, or lack of undoing, that was at the root of his oncoming health issues. Try as he might, MC could not halt the progression of his cancer by attempting affirmative and positive thinking. His attempts to forgive were not happening. His ability to stay in a state of silence was extremely difficult. When he became stimulated, his mind was activated into high gear. Silence was a thing to strive for but never seemed attainable. MC's effort to silence his mind was threatening to his ego's brain chatter. He always had the desire to rehash all his life's issues just one more time.

In one of his more austere moments, Sharinagar warned that one couldn't count on any man-made laws or science. He noted that science was inexact and that a certain altered state of wisdom should accompany it. Rudolf Steiner of the Anthroposophic movement referred to this as "spiritual science" and gave insight of it through his work, 'Knowledge of the Higher Worlds and Its Attainment".

For Sharinagar, science interpreted by the lower brain lead to the inventions of paranoiac and self-destructive tendencies. MC would later notice that this rationalism was prevalent in the cancer research science as well, which seemed threatened by intuitive and subtle science discoveries.

MC's interpretations of Sharinagar's warnings and forecasts did not always have the desired effect of quieting his mental chatter. Sometimes it seemed to exacerbate MC's lack of trust in life's issues and he would unwillingly continue to explore the doom and gloom sides of life.
Because of this, his emotionally-driven eating disorder was sometimes stimulated. The expression, "those who eat to live get healthy and those who live to eat get disease" was ringing true. The slightest disturbance caused MC's hand to go looking for something to put into his mouth. Certainly, MC could understand what Sharinagar was saying. However, at this point, applying the suggestions wasn't an option. While noted, they were put aside for future use.

Sharinagar was once asked if it was common and acceptable to fall into a passive state because of being unable to maintain silence. Sharinagar answered that this passive state was a state of self-deception.

Indifference is passive. Drugs and tranquilizers will create a passive state, but it is not energetic. The indifference of a passive state is not spirituality. This attitude often leads to clumsy and sloppy behavior and merely provides the pretense of being enlightened and aware. People in this state often learn the terminology and lingo of being self-aware and awakened using such words as Kundalini, Alpha state and spirituality. So-called gurus use this method to try to lure people who desire this presumptuous but phony peace. These leaders may even create discipleships to promote the sale of this passive, lethargic bliss, which is no more than recycled techniques and opinions.

Sharinagar went on to say that an enlightened state is alive, on fire, alert, giving all to receive all. A true voice comes from within and is shared with others in a compassionate way and without proprietorship. This living in the moment releases energy to the giver as well as the receiver.

"When the mind is still, it is energized. It will not tolerate deception, and it sees the false as the false. Those who withdraw do not know what real silence is. It is too energetic to be passive. If you cannot harness the energy of a truth then you only know the words."
Tara Singh, "Gifts from the Retreat"

Despite MC's holding dear to his past, he experienced some rather unusual breakthroughs in Sharinagar's presence. He would mention that there were two kinds of relationships: special relationships and holy relationships. When MC first knew Sharinagar, he was drawn to him as to an idol. Sharinagar seemed to do everything to discourage this but did seem to cater to some. To an onlooker, this appeared to be favoritism. MC's relationship with Sharinagar was unusual in that MC only observed but never judged him.

At that time, Sharinagar was always surrounded by people, most of whom wanted attention. They either had questions about personal problems or wanted to be personally recognized by Sharinagar.

During a retreat when it was impossible to approach Sharinagar, MC began to see the truth to what he often said, "The farther apart, the closer we are; for bodies get in the way at times."

MC had noticed people constantly looking for Sharinagar. They would literally stalk him, trying to get into special gatherings or encounters. MC observed a person peeking into a window where Sharinagar was staying.

Soon, MC realized that he was acting similarly. He made the sudden decision not to be part of this worshipping entourage. He became determined that he would simply line up at the end of the retreat to say his goodbyes to Sharinagar.

Shortly before the end of a retreat, an assistant of Sharinagar's approached MC and said, "Sharinagar wants to dine with you at the final lunch." MC had overcome a barrier and it was recognized. Sharinagar would always be there if MC had a need.

MC remembers talking to others about this and some admitted that they could not get over the need to have a personal acquaintance with Sharinagar. They would often drift away, looking for a more personal fulfillment. Many people would eventually fade from Sharinagar, especially those who wanted to learn or climb to some level of status too quickly. Sharinagar often frustrated them by saying it was nearly impossible for the brain alone to bring one to the determination and commitment needed for awakening. No one could do it for another; they would recognize it for themselves when it came. Since there was no prescribed curriculum to go about this awakening and no guaranteed apprenticeship to attain it, MC was left among those followers who were wondering when, if ever, this atonement or grace would occur.

It was around this time that MC was invited into a special gathering and was sitting near Sharinagar. As he was speaking, he made eye contact with MC while saying, "We need to turn all special relationships into holy relationships."

The next step in MC's relationship with Sharinagar came after a retreat when MC had helped him move some books. Sharinagar was winding down from the retreat. MC was due to take a plane home when Sharinagar asked him to come with him and a small group to his dwelling in the hills.

Before MC could even reply, Sharinagar said, "Don't tell me you will next time! Now is the time!"

The small group of cars started for the hills. Oddly enough, MC was given special treatment. Sharinagar insisted MC sit in the front seat while he sat in the back with another guest. During the ride, MC felt relaxed, comfortable, even blessed.

Sharinagar suddenly stopped talking in the back and leaned over the front seat addressing MC, "You see, this one is married and now her husband wants to leave and this one wants this and what to do....Do you see what hell it is to be a guru?"

Sharinagar and MC shopped together and ended up at a friend's house for dinner. Sharinagar was becoming far more casual than he was during retreats. It was the closest MC had been to what Sharinagar called being in the moment.

Sharinagar started sharing the difficulties that he had with his own children and how they behaved in peculiar ways, denouncing him to his friends, causing damage with some of those relationships. Hearing these common woes, there was perhaps a slight let down in MC's belief about Sharinagar's infallibility.

This triggered a voice in MC saying, "Oh, shut up!" He was reminded of a comic movie where a devil-like puppet sat on someone's shoulder offering its host sinister advice. MC panicked. What was this voice?

As Sharinagar said something to further expose his vulnerability, MC's rogue voice said, "Ya, ya, ya, what of it?" MC could not stop the voice; it was stronger than his will.

Things intensified when they left to go to Sharinagar's place.

"In the front," Sharinagar insisted.

Once in the car, he grabbed MC's shoulders from the back saying, "Tomorrow we'll go to this place where I once heard Krishnamurti speak."

Suspecting that Sharinagar could read people's minds, MC was now ill at ease. He was trying with all his might not to hear the contradictory imaginary voice with its indiscriminating statements. Suddenly, MC decided that he was in charge of these voices. The abrupt voices went dead. He realized the voice had manifested after he had passed judgment on Sharinagar. An internal duel had ensued.

Later that same evening, Sharinagar spoke of the dilemma resulting from the duality of man - the body with its conditioned brain versus the inner mind that imparts wisdom. The body knows only reaction with its conditioned brain while the inner mind knows only to respond.

MC thought later that it might be Sharinagar's lesson on the duality of the mind that he had experienced while struggling with the rogue voices in his mind. MC had never experienced such a phenomenon and it wasn't exactly what he thought of as a spiritual experience.

After that evening, Sharinagar seemed to trust MC and they began to travel together.

It seemed that all possible betrayal came upon MC that summer. After coming back from a trip to India, MC's bowel condition degenerated. No state of mind was safe from this feeling of vulnerability as MC began to bleed on a daily basis from his bowels. The cancer was beginning to cannibalize his body and toxins were affecting his behavior. Though MC had heard Sharinagar for years, he was still unable to forgive or apply any of these virtues to his own life. He mumbled to himself that often declared statement of Sharinagar's, "A truth once heard but not applied could turn to poison."

On the eve of Thanksgiving, MC collapsed in the emergency room of Valley Hospital. He would need surgery for a total blockage of the intestine. It would require a colostomy.
It was a makeshift plan since the probability of survival was minimal. MC was totally unprepared – and uninsured.

The Surgery

There was no time and yet time had stopped. It wasn't long. It wasn't quick. MC was no longer in charge of anything.

It was Thanksgiving Day and his surgeon Dr. Navaja had a difficult time finding anyone to assist him. Finally, he was ready.

Most of the people in the surgery unit knew MC personally. One could see the large intestine pushing through MC's abdomen. MC remembered well the colonoscopy. It was short and only had to go up a foot. Dr. Navaja thought that if he could go through the tumor's apple core center, he could release the pressure. When the scope got to the tumor, it looked like a large conch shell. It had a shiny pearl-like appearance in shades of pink. It was as if an octopus had been stuffed into a pipe. Dr. Navaja was astonished that MC could have let this go for so long. Then, while sighing, the doctor concluded that everybody waits too long.

As the doctor tried to release pressure, MC began to bleed. He flinched. He could see and feel the contact.

Dr. Navaja stopped, "I guess we are going to have to operate quickly. It would have made things a lot easier to have emptied out the colon."

He explained that since the top part of the colon was extended and the smaller bottom section was shrunk, he would have to put in the stoma.

As MC was rolling down the hall on a stretcher, he could see that everyone was busy in the operating room. They had all been called in from their turkey dinners. MC's arms extended sideways and were strapped to the stretcher.

MC said, "This looks a lot like the crucifixion."

The physician looked down at MC and said, "The crucifixion?" After a second, the doctor laughed.

That was the last thing MC remembered.

The next thing he remembered was a nurse asking him if he was in a lot of pain.

MC replied by saying, "A lot less pain than before."

MC's state of mind was changed. He now understood why a recovered alcoholic would say, after having been hopelessly addicted, "This is the best thing that could have happened to me!"

MC realized what Sharinagar meant when he asked of MC, "What is it going to take to make you decide?"

MC was accepting that the cancer was of his making.

Since it was Thanksgiving, the hospital wasn't that busy. MC was fortunate that the nurses had time for him. One nurse suggested that rather than calling the stoma a bag, MC should give it a more positive name. MC saw the need for positivity and decided to call it temp for temporary, which was for him the only hope.

The news of MC's surgery reached the foundation that cared for Sharinagar, now in his mid-eighties. MC desired to talk to him and Sharinagar called while MC lay in his hospital bed.

MC explained what had happened and what the outlook was. Sharinagar was in a state of receptivity where what he was about to say was not thought out. MC had observed this before in Sharinagar. His responses were not always linked to the problem or questions asked. Inadvertently, he would respond to the issue at hand.

He finally said, ""You seem like such a good person."

In lectures, Sharinagar had often stated that 'good' and 'bad' were meaningless terms in a meaningless world.

MC thought to himself, why would a good person get cancer?

MC was no longer banking on his body's strengths or rational thoughts. He went into a state of being completely receptive. Sharinagar suggested that MC contact his friend Dr. Beauve.

Sharinagar then said, "Where there is trust there is faith. Write that down."

MC would repeat the phrase while rolling down the hall at Valley Hospital when having his reconnecting surgery.

At night, a few days after the operation, MC pondered what he could decide and what he could trust in. Earlier that day, MC had read Lesson 136 from "A Course in Miracles" titled "Sickness is a defense against the truth."

At one point that evening while MC was talking to one of the night nurses, she stopped doing her monitoring and stared at MC in wide-eyed amazement. He asked her if she was bothered by what he was saying.

"No, not at all," she said.

MC had no exact recall of what he was saying, but he knew it had to do with a new understanding about his life. His terminal condition had changed the order of importance by which MC was going to live. He felt rather peaceful and content while this experienced nurse was staring at him so intently. He wished he could have remembered what he had said but she certainly must have been used to delusional patients in her experience!

That night seemed to MC a turning point. A determination seemed to have been born. By morning, the nurse would relate to MC that she had never had a night go by so fast.

In six months, MC would be well enough to attend the last retreat that Sharinagar would give. It was there that MC would ask him about breathing (pranayama) and would receive the greatest demonstrated lesson. It was not so much on how to do it, but why to do it. The techniques were of less concern than the need it fulfilled and the result it would bring.

MC would later take time to visit Sharinagar in his final days. After Sharinagar died, MC was recognized as one of the successfully treated cases of colon cancer listed on Dr. Mandel's enzyme therapy website. MC was relieved to know that one could bring the body back to health without having to be enlightened. Sharinagar talked about making contact with other forces. When ever-changing negative thoughts and past projections led to MC's failed health, he was left with being attentive, honest and determined to communicate with the inner mind.

During his time with Sharinagar, MC had picked up some important habits that would help him with the cancer treatment. Sharinagar had great admiration for the works of Edgar Cayce, the American sleeping prophet. Cayce suggested that people should always start the day off drinking a cup of hot water. This helped the liver out greatly in cleansing the body. Sharinagar had always known this from Ayurvedic traditions of India. Another suggestion of Cayce's that Sharinagar voiced was to eat a few raw almonds a day to prevent cancer.
Sharinagar would share his breakfast drink recipe with MC.

Put into a bowl to soak: three pitted dates, three stemmed figs, one TB sunflower seeds, 1 TB pumpkin seeds.
Alongside this, soak five raw almonds.

The next morning, peel the almonds, discard the peels and combine with the rest of the steep.
Combine this soak with a banana, strawberries, or any desired fruit into a blender to create a smooth creamy drink.

While Sharinagar was sharing this recipe, someone asked why one had to peel the almonds.

Sharinagar replied, smiling, "So the skins don't get caught in your teeth!"

Actually, the skins on the soaked almonds are enzyme inhibitors as well. MC combined this creamy drink with 14-grain gruel. The gruel consisted of equal amounts by weight of wheat berries, buckwheat, rye, barley, oat groats, millet, sesame seed, brown rice, flax seed, alfalfa seed, lentils, mung beans, corn and almonds. He ground two to four tablespoons of the 14 grains in a food mill and soaked that overnight in the refrigerator with as many tablespoons of water.

In the morning, this raw gruel was topped with 1 tablespoon of flax seed oil and 1/2 teaspoon of vital 10 powders, which is a pro-biotic blend of, among others, Lactobacillus and Bifidobacterium.

MC would then pour Sharinagar's creamy fruit and seed blend on top of the soaked raw grains which, when soaked overnight, had acquired the flavor of sprouting enzyme activation. This became a major part of MC's eating protocol that included eating raw enzyme-filled food to eliminate his cancer. For the first two years of the cancer treatment, MC ate this every day as his primary meal. It satisfied many of his requirements in an easy and delicious way. When MC combined this with his daily quart of raw carrot and vegetable juice and a daily leafy salad, he was well on his way to meeting the 70% raw diet required for his moderate vegetarian type of enzyme therapy prescribed to him by Dr. Mandel. MC needed to avoid anything that contained soy, which Dr. Mandel said was an enzyme inhibitor.

Sharinagar told MC that when one became aware that one is not just a body, one begins to take better care of the body. When given what it needs, the body acquires energy from its surroundings.

Once the commitment was made, MC seemed to be acquiring an order in life.

My voice never tells me anything; it only intimates to me when I deviate from the law.
Socrates

9
THE DECISION

Sharinagar often said that there was a big difference between a decision and a choice. A decision has certainty behind it and choice is ruled by fear, superstition, impulsiveness and learned knowledge. Choices are uncertain, more passive in nature, relating the past to the future. Decisions are inwardly guided; on fire and with more passion; in the here and now.

In MC's situation, the choices given him were dismal. The facts were that MC had a 5% chance of living two years if he did chemotherapy and less than a 5% chance of living two years if he did nothing. There had been some vague statement by Dr. Navaja about surviving six months if MC did not do chemotherapy. MC's son was told to say his farewells to MC the night before the surgery, for it might be the last chance. His son was also told that if MC made it through the surgery, he might have two months left to live. MC was spared the impact of these statements.

The definition of doing nothing does not take into account what can go on in the mind. To MC, the prognosis was doing nothing either way.

MC thought about some past decisions and how clear they had seemed, how they often seemed to fall into place easily. Even after making a decision, MC realized there was brain chatter at work. Inevitably, he would fall into a mode of feeling he needed to make the "right" choices to enhance his business, his health or his spirituality. MC now had to pay for the consequences of his choices. In his collapse, he could see that he had chosen a self-destructive path. Even with a brain full of theories and the tutelage of Sharinagar, MC could not avoid this abdominal tumor.

The external world does not acknowledge intuitive thinking. Often, more options are offered to compensate for poor choices.

Decision-making involves posing questions by using all of one's capacities. There is always a temptation to believe that someone else will make a decision for you and recommend some good options. Indeed they do, as MC's encounter with the oncologist Doctor Calvaria would demonstrate. MC now had to make a decision about what to do.

Sharinagar's friend Dr. Beauve came to mind. MC had never really been his patient but the doctor had offered him some guidance in the final days before the surgical removal of the tumor. Dr. Beauve's intuition complemented Sharinagar's. Their discussions always allowed the participation of the other.

Sharinagar would always say, "What is going to make you decide?"

When MC called Dr. Beauve's office, he said to the secretary, "I'd like to talk to Dr. Beauve about my treatment of colon cancer and how to go about it."

The secretary replied, "What kind of treatment do you want his help with, chemotherapy or something else?"

The word 'chemotherapy' shocked MC. He thought Dr. Beauve worked with alternative therapies. "Well, I wanted to know what steps to take."

The secretary said, "He will help you with whatever you decide to do."

When MC hung up the phone, he felt he was now really in crisis. Then, calmness overcame him. There seemed to be distractions along the way with many paths. To true healers, secretaries serve as triage for the undecided and save the doctors' time. They are not salesmen.

MC got back on the phone with the secretary and said," I would like to talk to Dr. Beauve about an alternative approach to dealing with my colon cancer. I have no interest in doing chemotherapy."

"Okay," said the secretary, "I will have him call you." MC had made a decision.

By now, a month after the colostomy, the surgeon, Dr. Navaja, suggested MC consult with the oncologist Dr. Calvaria. As Dr. Navaja knew, MC was stalling and leaning toward the idea of nontraditional treatment.
Dr. Navaja said, "Chemotherapy is the only known cure for colon cancer."
MC never got upset at Dr. Navaja. He was a surgeon and quite innocent about his convictions on cancer. The expression on his face showed concern that MC was wandering away from what was being offered.

Dr. Navaja was quite sincere about the effectiveness of chemotherapy treatment for colon cancer. "The odds are not good."
Nevertheless, Dr. Navaja pleaded with MC to see Dr. Calvaria.

Unlike Dr. Navaja, who could look right into MC's eyes, oncologist Dr. Calvaria avoided true eye contact. He looked as though he might have been a little sedated and came on strong as The Oncologist. Although he was aware of MC's reluctance toward chemotherapy, he suggested starting the treatment immediately. This being just a few days before the holidays, there seemed to be an urgency to get the treatment started.

MC was without medical insurance and told Dr. Calvaria he would not consider starting anything until the first of the year when he would be insured. MC then stated that he planned to have his colostomy reversed before considering any treatment.

Dr. Calvaria jumped in and said, "There will be a great advantage to doing the chemotherapy right away. The inside of the intestines have very few nerve endings and, because you have the bag, it won't burn your anus."

As he said this, he joined all the tips of the fingers on his right hand. "It will save you a lot of discomfort."

Just to end the meeting, MC again insisted that he would not consider anything until he was insured.

Dr. Calvaria came closer to MC, with his right hand cupped as if holding an imaginary bag of 5-flurouracil leucovorin, and said, "The chemo for colon cancer is only $20 a bag".

He reiterated the 5% chance of living two years if MC did chemotherapy. He reminded him he wouldn't have much time at all if he opted out of it.

MC was mentally drifting away and starting to plan his escape from the encounter.

Dr. Calvaria continued, "We need to get going. Tumors will move into the brain within six months if you don't treat it now."

Although not in Dr. Calvaria's web, MC thought about what anguish a statement like that would cause someone who was not aware of any other option. MC himself began to feel anguished despite his resolve.

Finally, the visit was ending. Dr. Calvaria had the disgruntled look of someone who missed the mark and could not make the sale.

The doctor was affiliated with Hilltop Hospital and visited the smaller hospitals in the area. One of them, Valley Hospital, had a small section to administer chemotherapy treatments. This close to the holidays the hospitals were not very busy.

MC would be reminded of the visit for a long time. Although the chemotherapy treatment would have only cost $20 a bag, that would not have included the administering of it. MC was surprised when he received a bill for several hundred dollars for his visit with Dr. Calvaria.

Dr. Calvaria had been referred to MC by Dr. Navaja and, in a good will gesture toward Dr. Navaja, MC had agreed to the appointment. Being billed several hundred dollars was far from what MC had expected. Hilltop Hospital would continue to send this bill, which created further anxiety in MC's life.

MC had not wanted Calvaria's sales pitch and he did not want the bill. MC's experience with Dr. Calvaria was further confirmation that he should take responsibility for his own treatment.

Sharinagar always said that a decision embodies more than just choice; that there is guidance separate from thought; that thought is incapable of making a real decision because decision is beyond thought; while choices are preferences that thought can justify. At best, a choice is going along with limited options. At worst, choice is being persuaded to go along with the only option.

To MC, taking charge and making a decision was tantamount to taking flight into the arms of a divine order. It needed to make sense and feel right to him. The prescribed right choice felt lonely, like an uncertain, indifferent compromise, like flipping a coin.

Sharinagar said often, "Once you have choices, you're lost."

Once MC had made the decision, he become alert, asked questions and became patient while waiting for inspiration from that calm voice within. He was less preoccupied with the outcome than with the action of the decision.

MC was often asked how he had found this Mandel method. It was several methods, actually, and it was more like recognizing them than finding them. When not made alone, a decision requires some discrimination. Simply "giving it up to God" might not be much different from handing yourself over to laboratory experimentations. This discrimination requires self-honesty, sincerity and determination to drop old habits and beliefs. The enlightened ones always said, "Know thyself."

MC had to be vigilant to that voice within. Although it might sometimes be contrary to his beliefs, it was answering questions already asked. Know how to ask for what you want. It was a humbling experience.

The highest form of human intelligence comes from a questioning, critical mind. When the hour is late, the mind, needs to drop all superficial character traits and focus on the task at hand. It took this cancer treatment to bring MC to the awareness of a decision.

10

LOLLIPOP: THE DECISION, PART 2

Lollipop would become a facilitator to MC in making his decision. She attended the Thanksgiving dinner during the time MC was having his colon cancer surgery and quickly took a special interest.

Lollipop herself had dealt with lung cancer. She had finally opted to have a lung surgically removed and had recovered nicely. Her diet undertook a drastic change. Her life had taken a different turn as she became quite savvy about cancer and a whiz at finding information on the computer. She took on gathering information for MC. In turn, MC confided in Lollipop about what he'd dubbed, *A Cancer Treatment*.

Though MC was considered brave to have taken a path so bold, he did not see it this way at all. To him, it was the chemotherapy patients who were courageous to undergo a dangerous and not so terribly effective treatment. MC lived in dread of the chemotherapy treatment suggested to him by his Valley Hospital surgeon Dr. Navaja.

MC thought that once people found out about the enzyme therapy, they would opt to do it instead. He felt so fortunate to have been guided to the cancer treatment he underwent that he was stunned there weren't mobs of people waiting at Dr. Heil Mittel Mandel's office. By MC's estimation, there should have been a queue of people going around the block near his office. Lollipop had been so involved with MC's treatment that she had become an advocate for Dr. Mandel's cancer treatment.

MC was rather respectful of authority but not totally trusting of it. When dealing with illness, MC noticed most people complied with the conventionally prescribed treatments. Even people who were rebellious in nature took exception when dealing with health issues. It was as if they had been handed a final verdict from their hospitals.

Because MC had witnessed unsuccessful cancer treatments within his family, he was grateful to find another way, despite having been told that he had a low probability of survival.

What happened to Popsy was well within the realm of conventional cancer treatment. When she first got lung cancer, she had considered treating it unconventionally. She changed her life style by undergoing a complete diet change and no longer smoked cigarettes or drank coffee. However, along with all of this she had the affected lung removed and then radiated. She bounced back rather well. This had all happened fourteen years before MC had his colon cancer surgery.

One day MC came across some statistics about lung cancer. He read that whether lung cancer was treated or not, the results and survival rates were about the same. This included surgical removal, chemo or radiation. Despite these statistics, most opt in favor of treatment over non-treatment.

What was further shocking was the high occurrence of mesothelioma years later in people who had been radiated near the chest. MC assumed that the rationale of conventional cancer treatment was to prolong life for a few years. The risk of greater problems down the road was looked at less intensely. Popsy seemed to have beaten the odds.

At about the time MC had been a four-year-survivor, Popsy felt less well. She had slowly altered her strict diet and was no longer doing her meditative exercises.

Popsy said to MC at one point, "I think it's back."

A doctor's visit did not seem to indicate any problems. Soon, imagery showed that Lollipop had mesothelioma inside the cavity left by the lung that had been surgically removed and radiated.

Popsy's son Deke had come to visit. While reading in the waiting room, he came across a brochure about mesothelioma. Deke started fretting as he read that on average, 18 years after radiation for lung cancer, patients would often acquire this form of cancer. It had been eighteen years.

Deke said, "Couldn't anyone see this coming? It's written right here in this pamphlet!"
Mesothelioma is a cancer that comes back with a vengeance. Treating it by conventional methods has

proven to be unsuccessful and does little to alter its progression.

G. Edward Griffin, author of "A World without Cancer: The Story of Vitamin B17", points out that an early detection often does not extend life but it does increase the duration of medical treatments a patient endures.

The fatality report and the statistics on cancer were something MC had experienced first-hand. The reaction of some people diagnosed with cancer would become the one perplexing behavior that MC could not fathom.

When it comes to authority, MC felt that he was rather docile and passive, maybe even too obedient at times. He never thought of himself as defiant or as a troublemaker. There was no rebelliousness behind his decision to do his cancer treatment.

At the start, MC was rather secretive about his decision. He no longer had the need to draw attention. His resentment, bitterness and blame would fade. In time, he would become detached from his cancer. MC noticed that some people who might be considered rebellious become submissive when deciding how to treat their cancer.

Popsy was informed enough on the subject of cancer to offer advice about it. Now she was calling MC with questions about Dr. Mandel and how to go about getting more information. She had been calling around to various Cancer Centers, as well. At one of these large institutions, she mentioned Dr. Mandel to a doctor.

After a moment of silence, the doctor replied, *"Mister* Mandel is a very nice man, but I'm afraid he knows nothing about cancer."

Popsy, referring to MC said, "But…but he's helped a friend of mine…"

The doctor said, "Your friend probably never had cancer in the first place."

Popsy relayed many such encounters to MC. She seemed to be distraught about the negativity surrounding Dr. Mandel.

MC guided Popsy through the preliminaries on how to get started with Dr. Mandel. He told her she had to insist that her medical records are sent to Dr. Mandel. Indeed, her hospital was hesitant to send these records to an "outsider." MC had had the same experience and warned Popsy to be persistent.

Popsy called MC and told him she was upset that Dr. Mandel was not returning her calls. She was scheduled for chemotherapy in the middle of the month, before the holiday, and needed a response from Dr. Mandel before then. Using the holidays as leverage to bump up treatment dates was, by now, a familiar scenario to MC.

MC phoned Dr. Mandel. The doctor did not seem to recognize Popsy's case.

He said, "Look, I need to see some medical records before talking to her on the phone. That would be totally unprofessional."

Puzzled, MC replied, "You have no records?"

Dr. Mandel reiterated the steps to be taken.

MC's confusion was cleared up after a call from Popsy that came soon after he had hung up with Dr. Mandel.

In tears, she said, "I've been rejected by Dr. Mandel!"

MC did not see how this was possible. He replied, "Even with a low chance of survival, Dr. Mandel will usually treat you as long as you insist."

Popsy said, "I wonder if my letter to Dr. Mandel was straddling the fence too much?"

MC was stunned at that and said, "Just cancel the chemotherapy appointment and I'll call the doctor back to find out what is going on."

Popsy said, "I…I can't cancel the appointment."

After hearing Popsy's hesitation, MC realized something: Popsy was looking for a way out of Dr. Mandel's treatment. She had only been appeasing MC in order not to disappoint him.

MC had never seen her letters and had been misled about Popsy's true intent.

He felt foolish and awkwardly told Popsy to do whatever pleased her

Popsy said, "I always feel obliged to do what my oncologist tells me to do and this is the way I want to go!"

Popsy's sister later told MC that their family had always felt they had good luck with conventional doctors. Their father had gone through a typical treatment for leukemia and died within the statistical timetable established for leukemia.

Popsy never seemed to question her choice to undergo chemotherapy. Her initial decision to remove the lung and radiate had come back to haunt her. She was trapped in cause and effect. Popsy seemed attached to her suffering. The cancer had become a part of her identity.

MC had limited contact with Popsy at this time. Her husband told him that following her first chemo treatment, Popsy had been found wandering around the hospital delirious and naked.

During this time, her husband was reading books on alternative cancer treatments. He continued to read them until her death, which came four months later. Ironically, a few years later, Popsy's husband read these same books during his own chemo treatments for urinary bladder cancer.

This episode with Popsy was MC's first introduction to learning of the struggle many go through in deciding how to treat their cancer, which seemed worlds apart from the ease in which MC had decided his course.

Dr. Mandel would explain why it was so important for him to screen patients. The reason was twofold – it was a way for the doctor to protect his practice and a way to help patients take charge of their treatment. Allowing patients this control was never permissible when undergoing conventional cancer treatment.

MC would talk to hundreds of people about Dr. Mandel's enzyme therapy. Only a few went all the way. The positive results were baffling to most but the cancer treatment had no sales pitch and no economic industrial sponsorship.

A nurse who had been diagnosed with cancer phoned MC for information. As soon as he began sharing, the nurse interrupted and rattled off information about Dr. Mandel and embryologist John Beard.

MC said, "You know more about this stuff than I do!"

He continued by telling her cancer was really a misfire of a system that was trying to help the body repair itself. There were similar mechanisms involved in these trophoblasts that manifested themselves within the body. One being able to hitch onto the uterine wall and duplicate an entire new member of its species, while other such trophoblasts latch onto irritated areas and are duplicating and regenerating organs and tissues.

MC went on, "It was recognized that malnutrition, dehydration, stress, drugs, chemicals, alcohol and various forms of radiation could alter trophoblasts and thus deform the fetus. Why would it be any different for the trophoblasts in charge of endoderm, mesoderm and ectoderm reproduction? The trophoblasts have the mechanism to create cancer. Dr. Mandel actually says the trophoblasts *are* cancer."

MC continued his layman's dissertation, "The hope of possibly understanding the origins of cancer may come from the observation of the fetus's development. From the original trophoblasts in which the fetus is formed, there is another formation of undifferentiated cells known as the placenta, which was observed to have properties resembling cancer cells. In fact, if the fetus died prematurely, the mother's body could be taken over by the placenta resulting in a cancer known as choriocarcinoma. The fetus feeds on the placenta. Starting on day 56 of its formation, the placenta cells begin to mature, which slows down their multiplication. That day coincides with the appearance of the pancreas in the fetus. With the help of the pancreatic proteolitic enzymes, the fetus will then digest the placenta for its nourishment. The amazing ability to digest the placenta becomes the key to eliminating cancer anywhere in the body."

The nurse was amazed. She concluded to MC that she was really interested in this "Mandelian method."

A short time after the call, though, MC heard the nurse had opted to do radiation therapy. The nurses and doctors treating her told her the radiation therapy was not working and chemotherapy would be pointless.

Nevertheless, she started being administered chemo even though all parties were fully aware of its futility. Much later, someone told MC that among the nurse's objections to the Mandelian method was its time-consuming procedures. The enemas, the juicing of vegetables and the packaging of hundreds of supplements a day were daunting tasks to her.

Another woman called MC and said she'd heard about his story. Someone in her family had been struck by the same cancer as MC's although it wasn't as far along.

MC started his well-rehearsed cancer story. He interjected that the decision to undergo Dr. Mandel's treatment was really up to the patient.

The woman said, "Well, it's my husband…well, he's right here…you might as well talk to him."
MC recounted his cancer story and the man was amazed. "What a story!" he said.
This statement puzzled MC. He replied, "Well, it could be your story, too."

MC began to think about his first phone call to Dr. Beauve. When the secretary had asked MC what kind of treatment he wanted to undergo, MC had panicked.

Yet, the direct question had forced him to declare his intentions.

MC tried this direct approach of questioning with the man on the phone. "What do you want to do?" asked MC.

The man replied, "Oh...uh...well, I'd like to try a combination of things."

MC was frank with the man and his wife, who was listening in on the conversation. He said, "You won't get through the screening process. The secretaries will protect the doctor from ever having to deal with your indecision." MC was convinced that this man would never go to see Dr. Mandel.

On his next visit to Dr. Mandel, MC mentioned the man who had called him. Dr. Mandel nodded and said, "I did see them. His wife mentioned talking to you. I wondered how they got through the screening."

MC realized that by telling the couple what not to say during the screening he had unwittingly enabled them access to the doctor.

He apologized to the doctor and said, "It was really the wife who wanted to do it."

Dr. Mandel nodded and said, "It was the wife who filled out the paper work. They probably won't follow through with it." He added, "Although it's important to have a supportive spouse, some can be so overbearing that it ends up being them rather than the patient making the decision. The patient definitely has to want to do this."

Dr. Mandel then shared a story. A woman and her daughter were having their initial meeting with him when, suddenly, the woman's husband barged into the office, pointed at Dr. Mandel and shouted, "How do we know you're not taking us for a ride and just taking our money?"

Dr. Mandel said, "I guess you have no way of knowing but it doesn't really matter because this meeting is over."
As the wife was sobbing, the daughter affirmed to Dr. Mandel that this behavior was a common occurrence in the couple's relationship.

Dr. Mandel told MC, "I can see why she got cancer in the first place."

All potential lawsuits come from this sort of family dynamic. MC had never recalled a single case where families would sue doctors or hospitals for horrendous results due to chemotherapy.

To MC, finding Dr. Mandel's method and being able to apply it to save his life seemed like an absolute gift. It was a bit more complicated for others. To some, the indecision could be tormenting. Adding to the anxiety is the oncologist's eagerness to begin treatment, which often forces an uninformed and rushed decision.

Of the people MC communicated with who were interested in blending alternative treatment along with chemotherapy and radiation, none tried the alternative route first. Alternative treatment would be tried only if chemo and radiation failed. Regardless of the gravity of their situation, they rationalized their preference toward chemo while complimenting MC on how courageous he was to have taken his route.

Ashley, who had known MC since her childhood, called him with great enthusiasm. She had heard about his cancer treatment and MC's success. Since she knew MC, she quickly informed herself about Dr. Mandel's enzyme therapy before she called MC. She explained to MC that she had been diagnosed with early stage breast cancer. She received chemotherapy treatment for a third time and the lump was still there.

Ashley said, "I'm really interested in doing the cancer treatment but I'm having a hard time giving up my safety net".

MC asked, "What is your safety net?"

"Chemo," Ashley replied.

MC was taken aback at how well imprinted her belief in the chemotherapy was. To consider a harsh treatment that had failed her three times in a row as a safety net was a telltale sign that Ashley would probably never end up at Dr. Mandel's.

There seemed to be a lack of trust in this rather simple approach to cancer. The philosophy of "kill or cure" seemed to be more believable. Despite its success, it seemed MC's case was thought of more as an exception, perhaps even a chancy and reckless decision.

It was Popsy's story as well as others that gave MC insight as to why there weren't throngs of people waiting to see Dr. Heil Mittel Mandel. His practice never seemed overwhelmed despite the doctor's television appearances and being written about in a best-selling book, "Knockout: Interviews with Doctors Who Are Curing Cancer – And How to Prevent Getting It in the First Place", by Suzanne Somers.

By MC's tenth cancer-free year, he had attended the funerals of many people who had shown interest in Dr. Mandel's cancer treatment but never had a chance to carry it out. Many had succumbed to complications and infections that seem to manifest during chemotherapy and radiation.

MC received a call from a school acquaintance who had previously spoken with MC about his cancer. This friend was interested in filming an interview with MC.

He told MC of Doug who had pancreatic cancer and was eager to talk with MC about treatments.

Doug came on the phone and told MC his story of chemo and various treatments he had undergone for his cancer. Doug ranted about his hardships and sounded fed up with it all. Now, the cancer was back.

Before starting his cancer story, MC explained to Doug that Dr. Mandel was renowned because of his success with some pancreatic cancer cases. Dr. Mandel was theoretically supposed to receive grants for a government research study on pancreatic cancer. Since pancreatic proteolitic enzymes are so vital in the elimination of cancer, it made sense that pancreatic cancer was considered one of the most dreaded.

To MC, pancreatic cancer was like playing soccer without a goalie. Taking pancreatic enzymes was essential.

MC suggested to Doug that he go down to Dr. Mandel and say, "I'm done with chemo. Been there. Done that. Now I want to do only the enzyme therapy."

After hesitating Doug said, "I guess I'm not there yet. I'm not at that stage quite yet."

He then started describing the nice doctors who had been working on him from the very beginning. They were now going to go at it with the "big guns".

"They're going to shrink the tumor."

MC recalled reading about an experiment where advanced cancer patients agreed to go through an experiment to see how much tumors would shrink using a particularly aggressive chemo treatment. Although shrinking was observed, no patients survived.

The faster the cancer is growing, the more rapidly it seems to respond to treatment, no matter what the treatment is. The tendency to go with the aggressive treatment for the aggressive cancers seems to override any consideration for spectacular recoveries using a science that jumpstarts an already existing immune system. It was assumed that MC's cancer could not have been serious to begin with because of its positive response to an alternative cancer treatment

MC remembers Dr. Calvaria's warning of the consequences of his advancing colon cancer and how urgent it was to start chemotherapy immediately. This alarmist discourse seemed to be the doctor's sole function as administrator of a small cancer clinic. All cancers in their final stages are aggressive and made more so by aggressive treatments.

MC was somewhat relieved at having spared himself the time and energy of describing his case to Doug, who seemed to want to get off the phone.

"Hey, thanks for talking to me," he said.

When confronted with the idea that this alternative cancer treatment would require an all out commitment, Doug's curiosity waned.

Popsy and her husband, although well versed in alternative terminology, took their places in line to be seen by an impersonal cancer industry that believes in the fatality of cancer. Popsy's will to live seemed diminished as the chemotherapy proceeded.

After her death, Popsy's husband often said during his own chemotherapy treatments, "The quicker it does its thing, the sooner I'll see Popsy."

In the obituaries of people who have died from cancer, it is often asked that all contributions be sent to the cancer clinic to further research development.

Some of these contributions were used by cancer institutes to supposedly reproduce Dr. Mandel's techniques in treating cancer. Their report concluded that Mandel's method did not warrant further consideration. Because of this, and in order to safeguard the vested interests of the cancer institutes, Dr. Mandel's promised grant was confiscated.

MC had always marveled at the control exerted over people in political and religious events. Popsy's example demonstrated to MC that the "war on cancer" promoted a similar control of peoples' belief systems and that this was an essential part of keeping the cancer industry alive.

The fact that MC's cancer case was extremely advanced and yet cured was proof that nutritional enzyme therapy was, to say the least, a viable treatment. MC referred to it as A Cancer Treatment that kept itself open to any and all science that could further the eradication of cancer while improving the health of body and mind. MC began to see himself as an escapee from a deadly failed state of oncological science.

Dr. Mandel referred to his office space as his home and to his treatment as the way to live. It required of MC a decision and the determination to follow a personal protocol of a science that MC, as a patient, could eventually comprehend. MC was grateful that the option to undergo A Cancer Treatment had presented itself to him. He was grateful that his determination to continue it persisted. MC felt that making this whole-hearted decision was a first step toward healing.

11
NEBUCHADNEZZAR THE CAT AND THE PANCREATIC ENZYME

Something that occurred during MC's recovery was an encounter with a cancerous cat.

Animals live by their instincts, not by their knowledge. Animal instincts have always amazed men who have somehow mistrusted their own instincts in favor of learned disciplines or preconceived notions. Empirical science seems to trust more in the experimentation on animals rather than having to deal with the psychosomatic disequilibrium of humans.

One of Dr. Mandel's patients who had heard about MC's cancer story, would end up doing enzyme therapy, not because of MC's story, but because he witnessed Nebuchadnezzar the cat's spontaneous response to pancreatic enzymes.

Nebud had developed cancer in the jaw. A large lump developed under his jawbone that would eventually abscess. The contour and inside of his mouth had turned a shiny black color. He became miserable and could hardly eat. Almost drooling, he was the picture of a miserable, dying cat. Nebud's owner knew MC and told him about this. She was aware that MC had dealt with cancer.

The visit to the veterinarian confirmed that Nebud had a malignant tumor, would be dead in a week and should be put to sleep on the spot. The owner did not want to put Nebud down and brought the cat home.

MC suggested that the cat take some of his pork pancreatic enzymes and gave the owner a bag to try out.

She gave Nebud 3 or four of the capsules each day. Each contained 425 mg of pork pancreatic enzymes. MC also gave Nebud's master a few of his 100 mg tablets of laetrile, (also known as amygdalin or vitamin B17), although he wasn't sure the cat would eat it since in its pure state it has a strong flavor. MC suggested that she wrap the pill in some cat food to see if he'd unknowingly eat it.

To the owner's surprise, Nebud became very interested in the pork pancreatic enzyme. She opened the capsules and dumped the contents directly on the cat food. Even though he had lately lost his appetite, he ate heartily. Afterwards, he even sniffed about looking for more.

At another of Nebud's feedings, the owner wrapped the small white pill of laetrile in some cat food. As Nebud started eating, the pill fell out of the food and onto the floor. He went to it immediately and ate it in one gulp. The owner was delighted. She had never been successful in giving Nebud any medicines no matter how well she disguised them. She was amazed that the cat seemed to know intuitively that these substances were good for him and was struck by Nebud's will to live. She continued to give him four capsules of the pork pancreatic enzymes a day.

The vet had lectured at length on the pain and suffering Nebud would endure if his life was prolonged. Two months after Nebud's diagnosis, his owner called the vet to assure her that things were going well. She explained that she had given the cat some enzymes that an acquaintance was taking for his cancer treatment.

The vet became very angry. "You should not be doing that," he said. "Products for humans should not be applied to cats."

Nebud's master quickly replied that these were pork enzymes and Nebud liked them.

Because of the vet's negative reaction and rather than deal with the scorn, Nebud's owner would later bring him to another veterinarian.

To understand what happened to Nebud during the first week after starting the pancreatic enzymes, it is necessary to go back a hundred years to the studies of John Beard, a Scottish embryologist.

By the year 1903, Beard had made a discovery about pancreatic enzymes. At the time, it was common knowledge that the placenta of a pregnant woman resembles cancer cells and is used as sample cells for cancer research. In fact, if a fetus died in-uteri, the placenta would sometimes grow out of control and completely invade the mother. This becomes a cancer known as choriocarcinoma.

Dr. Beard wondered why the cells of the placenta seemed to slow down and even stop growing during a normal pregnancy. It is at about day 56 of the pregnancy that the placenta halts its growth, which coincides with the appearance of the pancreas in the fetus. The fetus then nourishes itself by breaking down the placenta via the pancreatic enzymes. Dr. Beard postulated that excess pancreatic enzymes given to cancer victims would help dissolve foreign proteins and tumors.

MC realized that this discovery was the reason he, (and Nebuchadnezzar), were still alive.

To this simple explanation of pancreatic enzymes, we need to add another dimension: Pancreatic enzymes cause the cells to mature or differentiate. These mature, differentiated cells do not divide because as the cells mature, mitosis ends.

Nebud had a fast moving jaw cancer named feline oral squamous cell carcinoma. Flea collars and second-hand smoke can exacerbate this cancer. If detected at its initial stage, there is about a two-month life expectancy. When the symptoms of the tumor become obvious, the life expectancy is one to three weeks. Nebuchadnezzar was brought to the veterinarian at this later stage.

To understand how Nebud did so well so quickly, MC drifted back to Dr. Beard's explanation. MC imagined that as soon as the pancreatic enzymes reached the cancer cells, they would mature and stop multiplying. Could this have halted Nebud's bony cancer allowing him to keep going? Would it follow that his body would start dissolving existing tumors that fast?

In "One Man Alone: An Investigation of Nutrition, Cancer, and William Donald Kelley", Dr. Nicholas Gonzales says, "Kelley does agree with mainstream scientists that small numbers of cancer cells form in all of us each day, but only rarely do these mutants take hold and lead to clinical disease. Conventional researchers argue that the immune system, especially the natural killer cells, protect us from such malignancies. However, Dr. Kelley disagrees. He claims certain pancreatic enzymes, particularly the proteolytic or protein-digesting enzymes, and not the immune system, represent the first line of defense against malignancy."

Around this time, the swelling under Nebud's jaw quickly came to a head and ruptured. The vet had tried to lance the abscess but it broke open on its own shortly after starting the pancreatic enzymes.

Nebud seemed to be content and continued improving as time went on. The black shiny skin on the contour of his jaw retracted as well as the black inside his mouth. The abscess became small and grown over and did not seem to bother him at all. Nebud was happy and playful with the new cat, now partially grown, that had been given to his master to (needlessly) replace him.

Nebud had lived a hard life. He was acquired from a barn full of orange tabby cats, which had probably inbred for generations. As a result of his first three or four years as a Tom cat, one of his ears was crumpled up like tinfoil. He got sick at the age of four with distemper and his master successfully nursed him through that. Nebud was now neutered and around 13 years old. He was now a survivor of one cat year, which was about equal to MC's seven human years.

During the spring of this first year, Nebud started catching and eating mice.
An experiment was done with cats in which they were fed no raw food. It was observed that the third generation of cats on only cooked food could no longer reproduce, while acquiring social disorders such as cannibalism. They had a drastic increase in cancer deaths as well. The cats in this experiment were fed cooked food. It is known that most enzymes are destroyed by heating them to a boil or heating them in a microwave oven.

Here we need to state another peculiarity about pancreatic enzymes: They are not affected by heat or acid. Pancreatic enzymes seem to be exempt from these two hindrances. The embryologist John Beard injected pancreatic enzymes to dissolve cancers in patients as far back as 1903. He injected his patients with it because it was assumed that pancreatic enzymes orally ingested would not survive the digestive process as most enzymes break down with the acids in the gastro-intestinal track.

It is now known that the pancreatic enzymes do survive digestion and get to the cancer via the blood stream. Russian scientists have boiled pancreatic enzymes in hydrochloric acid for hours and through denaturation, were not able to extract the protein complex in the pancreatic enzymes. This phenomenon makes it possible for animals and humans to eat pancreatic enzymes from other animals and to benefit from the digestive enhancement for these foreign proteins intruding in the body.

Cancer, being just such a foreign protein, is dissolved with the help of other selective vitamins such as laetrile. The digestion of the cancer cells becomes a natural and specific function of these supplements. An intended response to combining pancreatic enzymes and vitamin B17 is to release arsenic into the foreign cell to destroy it, while leaving the rest of the body cells untouched.

As far as Nebud was concerned, he seemed to know instinctively what he needed. When placed in front of him, he consumed the laetrile immediately. Though animals operate via instinct, health plans and oncologists do not.

The myth that apricot and bitter almond pits are poisonous seems to have been promoted by a science that condones chemotherapy treatments. As far as chemotherapy goes, most body cells, as well as the immune system, suffer from the treatment.

When given pitted fruit, certain primates in zoos peel off the flesh of the fruit and discard it. They then smash the pit to get the almond shaped seed. As bitter as the seed may seem, they instinctively eat them first. In the wild, even lions, the king of beasts, will sometimes eat the digestive organs of their grazing victims before consuming the herbivore's flesh.

Only man, the most evolved creature on the planet, lives in fear of incurable diseases and stands in doubt about what to eat and what he needs. Yet Nebud, dependent on and domesticated by man, was able to recognize his cure when it was made available to him. Even in his cancerous state, he seemed to know what he needed.

Dr. Ernest T. Kreb is credited as being a pioneering laetrile researcher. At first, laetrile was described as working alongside enzymes and specific to attacking only cancer cells. It was thought that their metabolizing effects would be released on the foreign bodies, while leaving other healthy body cells alone. More attention was then given to laetrile, which could be isolated in crystalline form.
Even though it is a naturally occurring molecule found in almonds, apple seeds, blackberries and even some grasses, the F.D.A. forbade the prescribing of laetrile and warned consumers of its poisonous effects.

The presence of nontoxic laetrile is phasing out of the modern western diets. Ironically, this regulation helped to establish an underground market for laetrile, now made from apricot kernels and pressed into pills. Laetrile clinics in Mexico soon followed.

Meanwhile the enzymes, which are perhaps the most important factor in cancer breakdown, were left in the shadows since they are more difficult to manipulate commercially.

Many raw vegetable juices lose their enzymes minutes after their extraction. These enzymes are so elusive, even the F.D.A. would have a difficult time regulating them.

Nevertheless, the intravenous injection of pancreatic enzymes is no longer allowed in this country.

MC's doctor, Heil Mittel Mandel, who practices enzyme therapy, stated to MC that taking especially large amounts of laetrile, (9,000 mg per day for ten days), without having an immune system well-endowed with enzymes and vitamins might be somewhat intoxicating, a waste of money and ineffective. Nebud took the pancreatic enzymes first and on occasion was given 100 mg of laetrile.

What makes Nebud's case most interesting is its simplicity, which made for a fairly controlled experiment. Being a cat greatly reduces the psychological or manipulation factors. Nebud was quite independent, as are most cats. All that was needed was the tenacity of his owner to give him his chance.

At the time when Nebud was a survivor of one year and MC of seven years, there was media coverage about the parents of a child with cancer who were at odds over what kind of treatment their child would undergo. A tug-o-war ensued. The mother and child fled to Mexico and were followed by federal agents -- all for the right to pursue a non-traditional cancer treatment.

Nebud was spared this aspect of the cancer treatment. Nebud's owner barely had enough money to go to the veterinarian, much less for euthanasia or expensive chemotherapy treatment. There was not a lot of concern over the fate of an old tabby cat. Nebud could not be manipulated by the fear of dying of cancer. No psychological affirmations or placebo effects could be applied to a cat. He did not undergo any of the purges, cleanses, raw juices, coffee enemas or even piles of vitamins. As his appetite improved, Nebud was given some raw meat, the pork enzymes, vitamin B17, (a nitrilocide derived from fruit kernels), and nothing more!

More than a year after his diagnosis, Nebud is now dominated by the younger cat that was supposed to have replaced him. This younger cat has matured and is larger than Nebud. They are separated during feeding time because the younger cat is interested in the pancreatic enzyme powder. He always comes to clean out Nebud's dish after dinner.

Nebud comes to greet his owner's car when it enters the driveway. He often sits on her lap and purrs. Nebuchadnezzar the Cat is the embodiment of a cancer survivor. Alas, one year and two months after his diagnosis, Nebud went missing. The weather was hot and muggy when he suddenly returned tired and hungry. He had been closed up in a basement for three full days. During the next two weeks, Nebud's health steadily declined. He was taken to a veterinarian and put to sleep.

Afterwards, Nebud's owner wondered why no one was looking into this enzyme phenomenon. What could be accomplished if one had some background in the field and knew what they were doing?

12

FATHER FESTUS: THE STRESS FACTOR

Can this cancer treatment overcome even the most stressful conditions?

This is a question that MC would ask himself at the beginning of his recovery. Could one step back from life's emotionally-charged situations and find some resolve?

MC Mac had always wondered if he might not be challenged by life around the age of fifty. There was some superstition along with the early death of MC's mother at the age of fifty that often clouded his mind with doubt.

At this point, MC had invested all his funds into a construction project for his business. This made MC rather vulnerable to taking on more responsibility or becoming sick. In addition, around this point, a business partnership had fallen through. MC felt a slight resentment at this and felt as if his world was a lonely place filled with greed and selfishness. As the economy declined each year, these impressions seemed more pronounced.

Then, with no particular warning, MC one day saw a small bit of blood in his stool. Painless as it was, it would persist.

MC's temperament began to paint a portrait of a betrayed and ailing individual, despite continuing to attend Sharinagar's retreats. Sharinagar would tell MC during this time that he had never seen him so miserable. MC's gratitude seemed to be replaced by the feeling of oncoming doom. This sentiment he had observed and picked up from his mother. The behavior was propelled by example from his father Festus's flippant irresponsibility, a sort of benign neglect.

After the colon surgery for the removal of the malignant tumor, MC began to recognize that events in his life, good or bad, seemed to be playing out in a particular and challenging order. Getting cancer and the decision to recover from cancer could not be isolated from the situation in which MC was born. The real decision would be how long MC would keep rehashing old resentments. How long would it take him to break from a merry-go-round of family haunts?

Sharinagar often told MC that family relationships were among life's biggest challenges but also present the opportunities to learn forgiveness and help one overcome difficulties.

He advised MC by saying, "If you don't forgive in this lifetime, you will come back again with the same people until you do."

MC was given the opportunity to practice these lessons throughout the cancer treatment.

Twenty five years prior, after MC's mother died of a brain tumor, his father was married within six months to the recently widowed Gladice Grabbit. Festus, who appeared prosperous, had inherited some properties and lived a very active lifestyle.

The new wife, Gladice, would gradually come to the realization that Festus was not as prosperous as all that.

During a visit, Gladice's daughter Malice became acquainted with MC's cousin, Frick. As his family's favorite son, Frick acquired his parents' land. Soon, Frick and MC's stepsister Malice were married and a "Frick and Malice Farm" came about.

From then on, it seemed more difficult to visit with Frick as Malice took on the task of keeping him from socializing. Malice was openly antisocial and irritable, often turning red in the face for seemingly unknown reasons.

The Grabbits were grafted onto the Mac family and both Gladice and Malice commandeered their respective relationships and properties.

At about the age of seventy, Festus was diagnosed with prostate tumor, though none of the details was ever made clear to MC. Although he had already started some chemo treatments, Festus went with MC to a homeopathic physician to inquire about treating the cancer with shark cartilage and other alternative methods.

After the visit, the doctor told MC that although his father wanted to try alternative treatment, something in his determination seemed to be lacking. He added that Festus was a very easily-influenced character.

When MC asked Gladice how long Festus's chemo treatments would last Gladice replied, "Until the end." It turned out that this would not be true. Festus fought these treatments, postponed them and missed many appointments.

Despite this, the doctor always seemed to say, "We'll get back on track in another three to six months."

Later on, Festus was prescribed a capsule that was shunted into his arm and would dissolve slowly. Festus called it "Zero Growth." This is how Festus explained it all to MC, anyway.

Festus certainly complained about the chemo treatments he received. He said to MC once, "Don't ever do chemo. It poisons my guts and kills my joints. I'm never doing it again."

Somehow though, the next time MC called, Festus had done another chemo treatment and was complaining about his joints and guts burning. MC never quite understood the change in mind.

One day, in the presence of MC and Gladice, Festus complained about the chemotherapy. He said, "I'm not doing chemo again."

Gladice stated in a slow and monotonous voice, "Now, Fess, we have already talked about this. Why do we need to talk about it again?"

Eventually, Festus's chemotherapy did stop completely and care was focused on congestive heart failure that had manifested. It was as if the chemo treatments had been completely forgotten. Festus would maintain this level of health for several years, dodging his treatments and medications as much as possible.

The Land Grab

Ultimately, MC was beginning to see that Festus needed more attention. He was slowing down and disharmony had settled into his and Gladice's relationship. Always making the case for needing help and money, Gladice often mentioned to MC that she might not live at the homestead with Festus much longer.

One day, Festus brought up the subject of the homestead with MC. "You need to go see Gus and have him deed you the homestead. Never mind your brother Toulouse; he will never do anything with it."

A school friend of MC's, Gus was the lawyer who had written Festus's will. Although MC and Gus hadn't seen each other in years, Gus never charged Gladice and Festus much for his services because of his friendship with MC.

On a day that MC took Festus to the dentist, Festus said, "Go see Gus today. So MC went to visit Gus.

Gus took the phone off the hook and they talked and caught up with each other during Festus's dental appointment. Nothing about the homestead came up. After all, the will had already been written.
On the way home Festus was uneasy while MC was rather content. As MC was preparing to leave the homestead that day, Festus became talkative.

MC and Gladice were standing outside as Festus started in, "The Grabbit boys have taken an interest in the place. Now that they have property here, they helped clear out the shed and sweep off the porch."

MC snapped to attention. Festus then pointed to the hill and said, "They got the land at the top of the hill."

MC looked toward the hill and as he turned back, he saw that Gladice had Festus by the scruff of his shirt and was pulling him sideways. It was quite awkward as Festus, even though being tugged at, kept on talking.

Finally, Gladice let go of Festus, composed herself and said, "We had to do this because of the taxes."

MC was in shock. "You have to sell land to keep land. Why was I not told of tax problems?"

MC was facing Gladice. Festus, who had wanted to get this off his chest was somewhat relieved and went to sit down.

Gladice was in charge. Her monotonous droning and complaining over the past two years had been in anticipation of this exact outcome. By claiming that her two Grabbit boys always helped 'Fess' do things around the homestead, the compensation could now be justified. This was offset by the clear fact that nothing had ever been done and the homestead was uninsurable and about to be condemned. Now the land around the homestead was parceled away.

MC left the homestead in crisis. Already suffering from occasional nagging pain in the lower bowel, he was familiar with this disgruntled state. From this point on it became nearly constant. Try as he might, the subject of the homestead came to mind almost constantly and, with it, the burning in his gut.

MC was soon on the phone with Gus. As the information unraveled, MC was stunned.

Gus said, "There were actually three parcels of land subdivided and doled out to the Grabbits. I never thought Festus would have the interest to carry it out but he always seemed to get the paperwork done. They had to wait quite a while for some of these parcels. This has been going on for ten years. I didn't realize you weren't aware of their plans. I had no idea that these borders were so close to the house, much less that the house was used as the corner post for the boundary lines because I never went there. I never charged him very much because he was your father. You know, I should not be talking to you like this. Gladice signed the papers as well so that makes her my client. I think you need to get a lawyer to discuss this and I can forward all the information. I feel bad that I'm in the middle of this. A will means something at death; any deed made before Festus's death will supersede the will. Good luck."

It's not unusual in second marriages to have such conflicts. Gladice had always stated, "I have all my things in order and Festus has all his things in order."

In fact, Festus had no such structure. All his affairs were freelance, so to speak -- a pile of papers stuffed in his desk.

A trust had been made for the Grabbits that would go to Gladice's kids. She could benefit from the proceeds. One could add to it but not take out. This one-way flow was a handy storage for loose change.

MC and Toulouse both had a tendency not to respond when certain anomalies surfaced. Whether for personal health issues or for threats to their livelihood, responses were often lagging. Since MC and Toulouse did not seem to be after things, the Grabbits and the neighbors had the idea that, with Festus's flippant benevolence, things were up for grabs.

Toulouse and MC thought that in the end there would be only the homestead. On that summer day, MC realized that nothing had been spared.

MC took a turn for the worse. Why did the land division seem like such a betrayal to MC? It had been in the family seven or eight generations starting in the late 1700's. All the developable land had been carved out. Only a house and a barn in disrepair were left along with a marsh and a hill. Some people pointed out the possibility of the whole property being given away. That might have been easier. Try as he might MC was handcuffed to the property and the dilemma.

Sharinagar often said, "A king trembles in front of a man that wants nothing." Those who are attached suffer. MC assumed that the increase in his intestinal bleeding was due to this stressful situation.

Just before the holidays, MC paid a visit to the homestead. There was a septic problem at the homestead as well as with MC.

It was Festus's birthday. Gladice, in her persistent monotone, pointed out that Festus got upset when certain topics came up, hinting that one should avoid such topics. MC was not well and did not bring up any 'topics.' Festus was becoming deaf. Gladice could not see well. They both seemed to be aging fast during this time.

The event that would change MC's state of mind would be his collapse at Valley Hospital due to a complete colon blockage.

In the years leading up to this, Gladice had always made a point of telling everyone how well MC had done in his career - how prosperous and rich he had become. Gladice had even convinced Festus that Toulouse was well off.

Festus would say, "Toulouse has saved every dollar he ever earned."

When MC pointed out to Festus that Toulouse was living with very little means, Festus would respond with "Oh, yeah!"

It was obvious that Toulouse was not prosperous but the notion that he did not need much to live on prevailed.

MC did not think of himself as prosperous. What little prosperity MC had would soon vanish as he landed in Valley Hospital with no insurance.

The rumors that Gladice had spread about MC began to make sense when partitioning things according to people's "needs" commenced. Gladice had two grandchildren who were mentally handicapped and, therefore, needier. Due to that need, they were great recipients of Festus's gifts.

The news of MC's terminal condition was a relief to Gladice and the rest of the Grabbits. It quickly brought Gus and MC's lawyer together to discuss the land issues. The two lawyers were acquainted and decided to entrust the remainder of the land to a Mac family trust. This would bring closure to a potentially complicated inheritance proceeding, as MC needed to make his will.

The next few months were a turning point in MC's life. There would be two surgeries, two months apart, which were exhausting. Only a mind exempt from resentment and attachments could cope with such bad news. This detachment that MC could not manage of his own free will was forced on him due to this life-threatening calamity. It was as if there was a grace period when these ruminations and furies did not exist. After a while, MC became grateful for the cancer and the betrayals. There seemed to be a feeling of serenity during this first year.

Soon MC had more strength and took on more responsibilities. MC's brother Toulouse came to see MC and help take care of Festus, who was becoming more anxious and more desperate for money. Small things were disappearing from the homestead.

When MC would ask why, Festus would say, "I owe Gladice hundreds of thousands of dollars," Gladice did not like him to state this aloud in public.

It was at this stage that MC could see Festus needed to pay property taxes and that Gladice still collected his pensions. Since Toulouse was unable to help Festus manage funds, MC stepped in.

As MC's health improved, he began to inquire at the town office about the homestead property boundaries. The town planning commissioner, Mike, said there needed to be fifty feet of property around a building. MC went to a grievance hearing and stated that there needed to be a buffer zone around the house because the property line was against the house, which would make the property worthless. Since the garden and lawn sheds were no longer a part of the homestead, the value of the sheds needed to be deducted from the homestead and added to the Grabbits boys' subdivision lots.

The assessor began to have a puzzled look as officials often do when they don't have any details to offer.

Mike spoke up, "Can't you have the Grabbit boys give it back?"

MC replied, "I don't think so."

About the closed boundaries, Festus would later say, "We've been mowing that lawn for years. The Grabbit boys won't bother you."

Mike then inquired about Festus's antique Lincoln.

"Oh, one of the Grabbit boys has it", MC replied.

Mike said, "Oh, I see."

Another assessor concluded from the deed's description that the property line came through the barn. This further confused things. They finally agreed that they would deduct the property value of the two sheds that were on the front lawn since they were no longer with the house.

In a small town things take a while to register.

A few months later Mike called and said, "Festus gave a neighbor a piece of land for a quart of maple syrup."

Since this wasn't a subdivision, not much could be done. Mike added, "As for the Grabbit boys' subdivision, there might be a problem. I wish I had known he wanted to sell that Lincoln! And you need to tell Festus, 'YOU CAN'T OWE YOUR WIFE!'"

MC had composed a letter to the assessors inquiring how a house could be used as a property borderline. Couldn't this be changed? All the rights-of-way went through the homestead lawn and around the house.

Gus once asked, "You still have some lawn with that. When you look out the window you still see only your land, right?"

MC said, "No, Gus. When you look out the side and front windows you see only the Grabbits land."

Gus replied, "Oh, God, what a mess. I'm sorry about this."

MC's health improved during this period and he noticed that his outlook had become more positive. Yet, he still couldn't help but wonder why the property problems couldn't be solved.

Help would not come from Gladice. Her daughter Malice and the Grabbit boys were smug and sat back as Festus began to seem more and more like a fool.

Sharinagar often said, "Don't use the higher forces as bell boys; ask for clarity and then step back."

Finally, due to MC's inquiry, progress was made. A letter came from Mike:

Dear MC,

I am enclosing a copy of the zoning bylaw as it relates to the area the property is located in. This will give you the set-backs you need to correct the problem. Anyone doing what Festus did is required to have a site plan approved prior to subdividing their property. The reason for this is to prevent any nonconforming subdivision as has occurred in your case. I think you must seek legal advice and subdivide this property appropriately to make the lots conform to the zoning bylaw. For the Low Density Area, it may also need to be surveyed to accomplish this. I realize you didn't have a say in the subdivision and now the job of correcting the problem has become your problem. Please feel free to call.

Sincerely, Mike, Planning Commission Chair

MC sent this letter, his own letter and one from his lawyer asking that the land be restored to its original shape.

The lawyer asked MC," What chance do you think you have of getting the land back?"

MC said, "About 5 percent." This was the same percentage given him to survive cancer for one year.

The lawyer appreciated MC's modest estimation.

All this information was sent to the Grabbit boys and a neighbor. Toulouse, who was staying at the homestead, told MC that there was not the slightest sign of a letter having been received.

At that point, Festus had probably not been informed of the news.

MC thought the procedure would drag on with endless obstacles. All the deeds needed to be undone and redone before Festus passed on. All parties would have to agree on what, how and how much. MC would take on the legal fees as a gesture of good will. There had been a whole year of non-communication, preceded by ten years of secrecy and now, in a matter of a few days, it would all be in the open.

One day, Toulouse saw Malice racing into the homestead biting her lip, her face red as a beet. She explained to her mother that there was a problem. The neighbor who had received the letters was causing a stink.

More importantly, a friend of Frick and Malice's came to them and said, "The talk around town is that Gladice has given the homestead to her boys at the expense of Festus's boys."

One of her friends dropped by and repeated the very same gossip. Concerned with what town people thought of her, Gladice was in crisis. Within hours, she had left a message to have MC get back to her as soon as possible.

The message had come early in the morning. That same day, another phone call came. It was Sharinagar. He had never before called MC out of the blue. Sharinagar was aging and becoming more forgetful. The fact that he would remember MC and want to talk to him made the bizarre day even more so.

He said, "I have been thinking of you all day. How is your health? How are you doing?"

MC replied, "I'm doing very well. Our friend Dr. Beauve has helped me find a Dr. Mandel. I am following his enzyme therapy treatment. On this very day, however, I'm dealing with a property boundary dispute."

Sharinagar reflected and said, "Let them do the talking. Give them space. Be as wise as a serpent and as gentle as a dove."

His final words were, "Ask for help...we exist only in relationship. Make your blood relation's hearts glad."

MC reflected on this call. Just how good was his health? He was never sure. This family bickering can't be the best thing for cancer recovery. Despite the turbulence, MC was strong and improving.

In the past, when MC felt anxious, he ate; behavior he recognized in Festus. At times, MC had no recollection of even having eaten during these spells. Sharinagar often said that a stimulated mind craves gooey, gluey foods. With the help of Dr. Beauve, MC had made some diet restrictions.

The alkaline diet that MC was now following was the perfect balance and discipline to get through such events. The diet seemed to create a more mentally alert state. At this time, MC was juicing four pounds of wheat grass a week, drinking a quart of carrot juice a day and taking many vitamins that enhance this alkaline state of mind.

In the past, MC would have churned at such events and consumed starch, sugar, chocolate and cheeses to stay energized. MC had been told at an early age that chocolate would give him fortitude. The keeping of the new diet served as a focal point for his resolve. MC now had the strength to catch himself at bitter thoughts and realize that he could think in a different way. There was great resolve to stick to the diet no matter what the outcome. This moderate vegetarian diet was the right diet for MC.

MC called Gladice back that evening. She described the events as she perceived them and told MC in a righteous way that everything needed to be brought into the open and discussed. MC let Gladice talk. Much was being revealed about her. She was not comfortable at all.

MC replied that since everything relating to the homestead had been a 10-year secret, all the information he knew had been detailed in the certified letter.

"Yes, ha ha, I know." When in a dilemma, Gladice had a nervous way of laughing. It could be perceived as smug but now she was clearly struggling with how all of this had come about without her knowledge.

MC realized Gladice was dominant over Festus, who was not included in this conversation whatsoever. Now he realized something else was at stake. Gladice needed to control this situation and, somewhere along the line, she had lost control.

MC was feeling more like the observer. He followed Sharinagar's advice and let Gladice do most of the talking. It seemed that all her reproaches were describing what she was guilty of doing. She even caught herself doing that and laughed before she finished her sentences.

Suddenly, because she needed to be in control so badly, she said, "I'm going to undo all these deeds and put the homestead back the way it was before."

Technically, Festus needed to make the declaration and the Grabbit boys needed to give back the deeds. "We want nothing back for this," she concluded with forced virtuosity.

The paper work was completed quickly and the homestead property, (with the exception of the neighbor who had fleeced Festus of a small piece of land), was restored to its original state. As expected, the Grabbits accepted all offerings and back losses.

As he had experienced when awakening from anesthesia after his colostomy reversal, MC felt an impatience to get things done. He needed to take charge and no longer simply observe. The Grabbits' unjust and unfair maneuverings created resentment and acid thoughts. How could this be miraculously undone?

MC needed to disassociate with the pillaging of the past. It was a preoccupying daily struggle that was certainly not helping his cancer treatment.

After the homestead deed changed, things did not calm down at all for the Grabbits and Festus's life was made even more miserable.

The new deed, made into a MC Mac Family Trust, superseded the will. There had been a promise to Gladice that several acres of land be given her in a vague, undisclosed and undefined location. Gladice had erased all of this with the flick of a pen. She even made an appointment with a faraway lawyer to see if any of the changes could be reversed. Her behavior traumatized Festus and created anxiety.

Though Festus and Gladice were void from any ownership or responsibility of the homestead, they had life tenancy.

Gladice had implied for years that she would move out some day. Both MC and Toulouse were cognizant of this and accepting of its inevitability. This further exacerbated Festus's anxiety.

At this point, Festus had nothing left. His properties, jewelry, bonds and antique car were gone. An estate worth many thousands of dollars inherited by Festus and Gladice a few years back had somehow been shoveled into the Grabbits trust. Gladice removed Festus from their joint checking account. Festus by now would qualify for welfare. He was reduced to doing transactions by postal money orders.

There was a call from a forester asking MC if he knew about the lumbering going on at the homestead. This was a breach of an agreement with the state. MC drove to the homestead only to discover that Gladice had taken a great deal of lumber proceeds for a 'debt' Festus owed.

When MC proposed that this revenue should go toward taxes and insurance, Gladice declared, "I don't see why I should have to pay anything to stay in Festus's house." She of course had a point. As usual, all of this was well orchestrated.

MC was having financial difficulties at the time and this seemed to be particularly stressful. Apologizing to the state for his father's untimely harvest further complicated things. Gladice would later deny taking any benefits from the lumbering.

Soon after that, Gladice called a meeting on what to do about Festus. She invited a group to be there, which MC liked since there were always discrepancies to what Gladice said or had not said. She invited Bogie, Festus's army buddy and closest friend, her daughter Malice with the bright red cheeks, three nurses and Peter, the head of the county care for the aging.

Gladice started the meeting insinuating that Festus was suffering from Alzheimer's. She claimed he was burning pots on the electric stove and that his driver's license needed to be revoked. Then she declared that she hoped MC would be sharing in the care of Festus and informed Peter that MC's address would be used for all the arrangements.

While Gladice was talking, MC thought about when she fell twice in one day a few weeks ago, both times unable to get up without Festus's help. He also observed that Malice's function during the meeting was to do her best to shift the responsibility for Festus from her mother to MC.

Festus's friend Bogie spoke up, "I don't think Festus is that bad yet. I'm not going to undermine him by stealing his car keys or disabling the vehicles. He added, "MC, have you got power of attorney for Festus yet? What does Gladice think?"

Gladice seemed not to have heard the question. Bogie asked her directly in a louder voice if she thought MC should have Power of Attorney.

In front of the whole group Gladice replied, "Oh, yes."

So MC said he would swing by to see attorney Gus and have it done. The meeting did not seem to accomplish what Gladice had hoped.

On his way home, MC stopped to see Gus. MC told him of the group meeting and the decision that was made giving MC power of attorney over Festus.

Gus asked, "What does Gladice think of this?"

MC replied, "She agreed for me to do it. Call her up."

Gus said, "I don't solicit business."
It was decided that MC would phone Gladice and ask her to call Gus to arrange things.
Soon after the meeting, Peter from the county care for aging called MC.

"MC, I've been thinking about your father. I don't think he has Alzheimer's."

MC said, "I've wanted to call you about that but I'm glad you saw it for yourself. You see, there is a control issue here. Gladice wants to gobble up the pensions the minute they come in and charge me for all taxes, utilities, insurances and now home care as well. I have no idea about his affairs or about his health."

MC then related to Peter the partitioning of the homestead and the struggle to get it back.

Peter was silent for a moment and then said, "I know that Gladice does not want you to have Power of

Attorney. I thought it strange that she agreed to it at the meeting.

MC said, "Do you see my dilemma?"

"Oh, yes," Peter assured him.

Peter told MC to call him if he had any questions. Care for the aging is difficult enough as it is but when there is discord in the family, the caregivers tend to bow out. Peter was never heard from again.

Festus bounced back and Toulouse came to live with him and Gladice.

One day, Malice came to the homestead and declared that Festus "didn't look right" and should go to the hospital. She did this three times in one week. Toulouse, who was somewhat unassertive, was finally told by the hospital that there was nothing wrong with Festus and he should continue giving him his water reduction pills. Malice's response to this was that her mother could not get any rest with Festus around.

Festus finally did end up in the hospital with congestive heart failure symptoms. He had a way of looking and being very ill and then coming back to life. This time he seemed so weak it was clear the end was imminent. It was agreed by all that MC should quickly be given power of attorney.

On the very day that Gus was preparing these legal papers, Gladice and her boys came into the hospital with pen and paper and suggested to Festus that Gladice could take care of the bills if she got the money from his account.

Toulouse, an observer and noninterventionist, had been in charge of Festus's 'debt' payments and felt compelled to find out more.

Toulouse came into Festus's hospital room in time to see the Grabbit boys holding a pen and paper as they propped up Festus while he attempted to write. He had an IV in his writing arm, but with the boys' assistance, Festus was able to give Gladice permission to empty his account. Actually, the account was in Festus and MC's names.

Toulouse finally approached the bed and said, "Uh, there are quite a few fuel bills and dental bills to pay off."

Festus replied, "Oh don't be so conservative," and sank back on his bed in exhaustion.

Toulouse witnessed Gladice and her two boys shuffling into the bank. One son helped her walk because of her bad hip. The other was there to carry out the thousands in cash. Festus's note had mentioned a transfer but the note was hard to read so the Grabbits replied that they preferred cash. Since the people at the bank, recognized Festus's handwriting, they obliged.

As Toulouse and MC reflected on these events they realized the Grabbits had taken advantage of their unsuspicious tendencies. Though not intellectual or even educated, the Grabbits were savvy about the ins and outs of legal proceedings. The family's shady behavior often left Toulouse and MC wondering if there were rules or safeguards against some of their activities.

"They can do that?" MC would ask Gus.
However, the Grabbits usually managed to remain a half inch within the law.

MC had become Power of Attorney for Festus. Resurrected once more, Festus was brought back to the homestead, as MC had promised him.

Gladice was preparing for hip surgery. She went into surgery leaving Festus and Toulouse at the homestead. When the appetite for life is dependent on another, one seems to be waiting constantly. Festus was a lonely man. All concerns about money, food or projects seemed to disappear when Gladice was gone. Festus would end up back in the hospital, which weakened him more. What gave him courage was that he and Gladice would be in the same rehab facility when he left the hospital. Soon, MC was delivering Festus to the rehab.

Festus was put in a room for two with an empty bed next to him. Married couples were allowed to room together.

Festus told Gladice, "Bring your stuff down here quick and claim it."

"Fess, I think it would be best if we stayed in separate rooms." She turned to MC and said, "Don't you think that would be best, MC?"

MC replied, "Gladice, do what you want to do, don't involve me."

The nurses joked how it was good for Festus to get up and walk to the opposite end of the building to see his wife. "It keeps him going," they said.
MC was abroad while Toulouse stayed at the homestead during Festus's recuperation.
Soon, events took a slight turn. Gladice was leaving the rehab and moving into an apartment, leaving Festus behind.

Since the Grabbit boys were aware that MC was away, they began to scout out the homestead. Toulouse contacted MC to apprise him of the situation. He told

MC the Grabbits were taking things from the house declaring that Festus had given them the okay.

MC called Festus's friend Bogie and asked if he would help mediate. When Bogie realized that MC was out of the country he agreed.

Bogie said, "Festus is alive! What are they doing?"

MC replied, "The lawyer said that if we carried out the will it would be a lot easier now than later."

Bogie said, "But a will is for when someone dies!"

Bogie went to the house and watched as the Grabbits took 'their mother's stuff,' that included the television and vacuum cleaner among other things.

"Festus said we could have it," they told Bogie.

Bogie shook his head while thinking that Toulouse gave in and let them take too many things. It was finally agreed that the rest would be settled when MC got back.

Gladice and the Grabbits always seemed to have well-orchestrated plans. As a result, Festus and the Mac boys always seemed to be waiting for the next performance to come flying down the homestead driveway.

At this point, MC had survived the 5% chance of living two years by a year and a half. If stress is truly a factor in the occurrence of cancer, MC was living proof that this cancer treatment could defeat the greatest of conflicts.

Back from his trip, MC attended meetings with his son, Toulouse, Frick and Malice. Toulouse would always call Malice 'the viper.'

Malice would often say, "Festus would have lost everything if it had not been for my mother."

At this point, Festus was abandoned at the rehab with absolutely nothing to his name and on welfare.

Frick spoke up and wanted to explain why Festus had lost his other properties. MC cut him off. Toulouse later told MC that he would liked to have heard Frick's justifications. MC knew it would only be another aggravation.

Basically, Frick and Malice wanted more things and it was agreed that these items would be placed on the porch.

At first Malice resisted, saying that the house was her mother's domain. Since she had overlooked that Festus was still alive, Gus the attorney quickly corrected her.

Malice finally conceded that her brothers were ignorant to have said they owned everything and they should never have taken the homestead land.

One would think this would have appeased Malice and the Grabbit boys. Except for the homestead, they had gotten everything they wanted. Yet they still appeared to be in a warring state of mind.

The most interesting thing to come out of this meeting was when Frick started showing some 'concern' about Festus.

MC had been inquiring about how to get Festus released from rehab.

Frick, slightly irritated said, "You know, Festus gets a lot of good care at that rehab."
MC was puzzled by this new founded concern and amazed at how news traveled so fast.

The next event came to light through Festus's army buddy Bogie. He was driving past the homestead and saw Malice's car followed by a pickup driving into the property. He decided to check and see what was going on. He recognized the pickup as one that belonged to a junkyard antique dealer from the next town over. He climbed the porch of the homestead and saw Malice inside the house with the junk collector. The sight of Bogie brought Malice out immediately.

Bogie asked, "What's going on?"

Malice replied, "Why don't you mind your own goddamn business?"

Bogie, a small man in his late eighties with a heart condition was dejected by this experience. He left the porch and drove off.

Bogie first told MC this story in a regretful way. However, soon he was so angry over the incident that MC felt compelled to interject.

"Look, Bogie, don't make yourself sick over this. It is our problem, not yours."

Bogie replied, "This has nothing to do with you. Gladice, that bitch, had me fooled. She probably will deny that her daughter ever spoke to me that way. All her lying about Festus and you boys - I now have a taste of how they operate." Bogie would never speak to Gladice again.

After this episode, Toulouse and Bogie fortified all the entrances from inside the house and padlocked the outside of the front door. There were still a few things belonging to the Grabbits on the porch. MC and Toulouse hoped that all these things would soon be gone and that everything would calm down.

Toulouse later went back to the homestead and noticed all the things on the porch were gone. However, the key to the padlock was not working.

In the back of his mind, Toulouse had always thought the Grabbits capable of locking him out of his own home. In anticipation of that, Toulouse had left a window open in the upper floor of the homestead. He climbed into the house with a wooden ladder he had placed near the window.

The local sheriff that stopped by told MC he had never investigated a break in where the thieves replaced the burglarized lock with a new one.

The sheriff said he would call the Grabbit boys but that this smelled of domestic bickering.

Pretending to be concerned, their mother Gladice later asked Toulouse if the sheriff had found out who broke in. Toulouse, who was of a kind nature, replied with a smile, "It's still under investigation."

MC now had to turn his attention to Festus, who had been labeled a long term care patient. He had to decide whether to leave Festus in rehab or to take him home.

Sharinagar always conveyed that the care of one's parents was an auspicious thing and often said, "Don't neglect your parents."

MC came to the rehab and saw Festus sitting with a stunned look on his face, a common expression of patients at the rehab.

"How are you doing?" MC asked.

Festus said, "Well, Gladice has opted to go it alone so I have no place to go."

MC replied, "I'm going to bring you home with me."

Festus said, "All I need is a small room somewhere."

"I have that for you at my home."

"Am I going to be cold?"

This last question struck MC and he discovered that Gladice had told Festus he would always be cold unless he stayed at the rehab. Festus had poor circulation by now and dreaded being cold. That very day, they were blessed with the first heat wave of the year.

Taking Festus out of rehab was a relief that outweighed any possible problems. MC did not have any concrete plans but once the decision was made, the doors of opportunity seemed to open up one after another. It was similar to the decision about the cancer treatment. Every single person MC met in those few days would be a link to carrying this out.

MC's next challenge was to untangle the affairs of Festus and his wife. In order to qualify for benefits, the state had to make sure that Festus was no longer associated with the "Grabbit Trust", which Gladice had handed over to her kids for just a long enough period to be disassociated from it.

MC's decision to take Festus home was definitely not on Gladice's agenda. It was now making sense why Frick was trying to convince MC to leave Festus in the rehab. Gladice must have been banking that Festus would not be healthy enough to go anywhere. Having MC take care of Festus was putting a monkey wrench in her plans. Festus's condition improved just on the news that he was leaving the rehab.

As long as Festus was in rehab all seemed normal. After he had been released, the question on the lips of nurses, doctors and state employees was, "Where is his wife?"

It was Festus's new doctor that introduced MC to the meaning of the word estranged. With the new scenario

making her look suspicious, Gladice started rumors that MC had taken Festus away from her and also driven her out of the homestead by asking for rent.
This was probably a defensive reflex to explain why Festus and she were no longer together.

She did manage to persuade some of the Macs in Frick's family to side against MC. He was always amazed at the influence Gladice exerted over these people. It turns out this was yet another scenario she had been churning for years.

What followed was a phone call between MC and Gladice. During this call, MC said, "Except for taking Festus home with me, it seems your plans are on schedule. I'll make sure Festus gets to visit you a lot. What's this about me asking you for rent? It seems to me you've been living off the land."

He mentioned the disappearances of properties, bank accounts, antique cars and lumber money.

"At one point, the front lawn of the homestead belonged to your sons!" exclaimed MC.

Gladice droned, "They did not own the front lawn."

MC said, "They tell me this is an example of being estranged?"

Gladice cried, "We aren't estranged! That makes me mad!"

MC replied, "Well, I guess I'm taking care of Festus?"

"It's about time!" she said.

MC was struck by the frankness of this statement revealing her strategy.
He continued, "Well, you have made out very well."

"Now I'm really mad," Gladice said. "You have made my life miserable."

MC was surprised – these were his exact sentiments towards her.

After the phone call, MC told himself, "People choose to be miserable or not."
That was MC's last real communication with Gladice. From then on, Toulouse would be the one who would take Festus to and from Gladice's.

Within her circle of friends, Gladice had painted a nasty image of MC driving her away and taking Festus from her. However, to the general population in the community, she told a different story – one in which she was in constant contact with Festus and that she was monitoring his care.

MC immediately started giving Festus some of the supplements and enzymes that he was taking from Dr. Mandel. Since these raw vitamins were capsulated, they could be pulled apart and the powder poured into a glass of juice. Hard, compacted and cooked vitamins were less effective and would have been an ordeal for Festus to swallow every morning.

Festus responded to a new diet very well. His fingernails began to grow out clearer and smoother. His complaints of arthritis were diminished, possibly by the intake of Type II chicken collagen. Soon he was practically off medications altogether. MC was using the placebo effect to accomplish some of Festus's health improvements. A visiting nurse was able to encourage Festus to circle the grounds at MC's with the help of a walker.
The doctor told MC, "You've accomplished miracles here but there will be a limit to how much Festus will recover. Because of the congestive heart failure, you

have to be vigilant for any signs of fluid around his heart and lungs."

MC resisted such forecasts just as he had resisted his own terminal diagnosis of colon cancer. The doctor's flattery made MC question his own motives. Being in charge of Festus had reduced a great deal of stress. He no longer felt he was being held for ransom, so to speak, by his father's declining health. However, was bringing Festus back to health motivated by the inconvenience it was causing Gladice and the Grabbits or was it an act of gratitude?

Festus was never strengthened by his visits to Gladice. Toulouse wondered if she might be slipping him something and began to stay near Festus during the visits.

Since Toulouse and MC's mother's valuables were missing, Toulouse began to obsess over what might be in some trunks in Gladice's apartment. Gladice kept an eye on Toulouse and never left him and Festus alone. Needless to say, the visits between Gladice and Festus's son were awkward. Gladice had been publicly critical of the plan to have Festus live with MC. She had proclaimed that she would never be able to see Fess. But now that Festus was mobile and brought for frequent visits, Gladice began to fabricate reasons Festus could not come.

"I have a hair appointment tomorrow, so Festus can't come."

Festus would pipe up and say, "That's OK, I'll wait. When is it?"

Gladice would say, "No, Fess, you can't be here when I'm not here. Rules of the building." Although there were no such rules.

"I'll come when you're done."

"No, Fess."

Perhaps Festus was counting on Gladice for his salvation. He looked like a small, abandoned puppy watching her every move, while confined mostly to a chair.

"I just can't take care of you, Fess," she would drone, especially when Toulouse could overhear her.

During his stay at MC's, he overheard Festus talking. "I'll be okay mama." (Gladice) "I got a room on Spoon St. and nurses taking good care of me."

MC said, "What are you saying, Dad?"

Festus replied, "Oh, don't mind me. I'm just thinking out loud."

There was a house for the aged located on that street. It was probably just another promise made by Gladice but never kept.

On another occasion, when talking to a case coordinator, Festus said, "My hope is to get a room with two beds at the county home for me and my wife."

The coordinator replied, "But you are getting excellent care here with your son."

Festus replied, "Oh, yeah, it's just an aspiration I have."

The care coordinator, who knew his situation, gave a sad sort of smile.

This had been the hard part for Festus. For the first few months after improving, he had wanted to go back with Gladice. Gladice apparently had no other recourse but to blame the separation on MC.

One day, on the phone with MC, Festus made a stand and declared that he was moving back in with her.

At first, Gladice didn't seem opposed.

"Transfer my accounts," Festus announced.

MC said, "Is there a bed for you? Are you wanted there? Sure you can move but is there a guarantee she won't have you back in rehab in two days?"

Gladice was heard in the background saying, "You'd better go back with MC, Fess."

Festus was becoming exhausted. Speaking softly so Gladice couldn't hear him, MC said to Festus, "Why do you think you are living with me if she really wanted you?"

Stumped, Festus said, "I'm sick, god dammit, I'm sick. I've got diarrhea. I can't do anything."

Festus had lost control of his life and, because of his ill health was being ostracized. He was caught in a situation between his sons who took care of him, and his wife whom he was still drawn to, even though she was out of reach emotionally and monetarily.

Toulouse mentioned to MC that occasionally on their way out from visiting Gladice, Festus pleaded with him to lay him down in the hallway next to Gladice's door for the night. This way, he could be close to her apartment the next day because Festus loved to have an early breakfast with her. Toulouse sighed but never replied.

Christmas was coming and Festus was looking forward to being at Gladice's for the holiday. Gladice had other plans. She planned a small pre-Christmas celebration for Festus and then she would be with her family for Christmas week.

Festus was not accepting this plan but his two sons were no longer interested in being mediators.

Since Gladice had left, Festus had absolutely no concern for money. Just before his pre-Christmas party, the topic of money came up.

"Where's my wallet?"

Festus was still in debt since having his bank account sacked a few months prior yet he inquired about his money.

Toulouse said, "You'll have to wait a few days, it hasn't come in yet."

Festus asked, "Can I borrow a few hundred dollars? I want to give Gladice money for her kids' Christmas clubs."

When Festus saw both his sons stiffen he added with a stammer, "And … and for her eye drops."

Childhood memories flooded MC and Toulouse. As kids, their time, their labor, even their money would be donated to further the cause of someone else or to try to impress someone else. Festus's manner of giving had never secured any respect from the recipients or from his two bewildered sons.

This current situation was the result of a tug-o-war within Festus between his worthiness and guilt. This struggle was one that Festus had passed on to both his sons.

His being dumped had probably been the result of these causes and effects. Festus's desperation exhausted both Toulouse and especially MC, whose care for Festus, the homestead and the cancer treatment costs had strapped him financially. Toulouse and MC were committed to care for their father despite the sometimes heavy baggage of a lifetime. But it was difficult for them to see their father exploited.

The Christmas club loans would have had no effect on Gladice. She was establishing a new precedent. A quick pre-Christmas evening was scheduled for Festus, who received a watch. Toulouse was ordered to take Festus away at a certain time.

Festus became very excited, almost agitated, at this party and talked nonstop as it drew to a close.

"I'll see you next week, Fess," said Gladice.

Festus had become subdued and then exhausted. A small glimmer of light, (hope?), went out that Christmas, which turned out to be Festus's last. From then on, he would complain of the long trips to-and-from Gladice's. His heart was broken.

MC's son Marco found an opened letter at the homestead that he thought might be important for the Grabbits to have. It was addressed to Gladice and she had written on the envelope, "correspondence file." Marco was trying to prove to MC that the Grabbits could be dealt with in a civil manner. The Grabbits never returned any of his calls, so Marco opened it up to see for himself. It was a letter from Gladice's granddaughter, Missle Toes.

"Dear Grandma,

I am glad, and I fully support your reason for choosing a smaller, more manageable place to live (one single bed.) Your health and wellbeing are important to your whole family, and this is a step in the right direction. It's unfortunate that Grandpa Fess doesn't contribute to keeping both of you healthy and relaxed as it should during retirement, but I guess that can't be helped at this point. He's still very dear to me; I just wish he would do what's best for both of you."

Missle Toes

Almost everyone agreed with MC that this was a document to keep. MC had always had a worry in the back of his mind that the Grabbits would come after him for compensation, especially after MC had been accused of driving Gladice from the homestead. Now there was an open declaration of Gladice's intentions written a whole year before. It was written before the emptying of bank accounts, before furniture and tools started disappearing and before supposedly being asked for rent.

It was of great comfort to MC while he took care of his father to have this written, candid insight into the Grabbits' plans. It was perhaps a delusional sense of security.

MC threw a birthday party for Festus. Macs from all over came to celebrate. Festus was able to walk around and visit with the guests.

At the last minute, Gladice opted not to come. Festus's buddy Bogie confided to MC that, if Gladice had come, he was going to "let her have it."

It was around this time that MC helped Festus with a liver cleanse; a simple protocol of drinking four cups of apple juice a day mixed with Phosfood liquid. Festus drank four cups of this mix four days in a row, along with his regular food. On the fifth day he ended the cleanse by drinking Epsom salts followed by whipped cream and berries for dinner. Just before going to bed, Festus took a half cup of olive oil mixed in grapefruit juice while lying down on his right side.

This cleanse had been one of the turning points for MC during the first few months of his cancer treatment. It had helped in ridding MC of gallstones.

In Festus's case, the visiting nurses had observed some talkative spells and more hyperactive behavior, but in a few days, they admitted there was great improvement.

The medical professionals react skeptically to alternative procedures that accomplish what they consider impossible results. They believe that the only way to remove gallstones is surgically. MC felt odd enough explaining his diet. It seemed to set him apart even more when explaining cleanses and fasts.

Festus had a good year, but there was a gradual decline. His mind was absolutely clear. His memory of facts, figures and dates amazed MC. It was sometimes aggravating to hear precise details of his benevolence. For the first time in MC's life, he would hear him talk about World War II. MC had often heard complaints from Gladice on how the Grabbit boys were tired of hearing Festus's war stories. MC had never heard a single one.

In an almost uninvolved but profound depiction of what he had witnessed, Festus described the events humbly and soberly, claiming no glory or heroism.

At one point Festus said to MC, "I hope I'm not boring you?"

MC remembers being in this disembodied state after his first surgery. He also remembers asking the spellbound night nurse if she was bored with what he was saying. She had told MC that it was one of the most energized and quickest night shifts that she could remember. MC later could not remember what he had said to the nurse. He could not remember much of Festus's war descriptions either but there was an uplifting transfer of energy between the speaker and the listener. Sharinagar described these as moments of awareness where time stops.

After about a year and a half with MC, Festus began to focus on MC for all his care and needs. Festus was failing.

It was at this time that MC got caught in one of his childhood habits and started eating chocolate. MC's doctor Heil Mittel Mandel later told MC that it was an impulsive attempt to get more energy to deal with a difficult period. In fact, MC consumed a pound or so during a week. MC was also starting his liver cleanse, which was on his scheduled protocol. He never broke his schedule. He just added undesirable foods.

MC proposed to Festus that they go to the homestead.

"What for?" asked Festus.

"To go visit … people", MC said, with Gladice in mind.

"I've seen enough people," Festus replied, but agreed to MC's plan.

While Festus was sitting in front of a window at the homestead the next day, he exclaimed, "There's my dog!"

"Which dog?" MC asked.

Festus declared that it was Tipper, his childhood dog.

MC, a bit concerned, asked Festus if he would like to lie down in bed. Festus replied, "Not yet, I'm watching the procession of things going by outside."

The next morning MC was to take Festus to the Veterans Hospital for a three-day visit. This worked out especially well as Festus loved going to the V.A. hospital. This would give MC a chance to prepare a more intense plan for Festus's care.

While driving to the hospital, Festus gently tapped MC on the knee and said, "You done good, MC, you done good." Festus left his hand on MC's knee for a while. Even though it was hot in the pickup, Festus's hand felt cold on MC's knee.

Then Festus said, "When this is over, MC, you'll get your settlement."

The cynic in MC pondered the choice of the word 'settlement.' MC interpreted this in a couple of different ways as he smiled inwardly but then he took it as a simple declaration of gratitude.

People often remarked on the great deal of love MC must feel towards his father in order to care for him. This seemed to be overshadowed by the antics of the resentful and spiteful Grabbit family. At that moment on a beautiful fall day, MC was enjoying a most pleasant ride with his father.

At the hospital, MC parked the truck, unloaded the wheelchair and prepared to transfer Festus from the truck to the wheelchair, something MC had done a thousand times by now. This time, as Festus was coming out the door he did not land squarely on his feet and, unbalanced, pivoted toward the wheelchair. Festus was in mid-air by the time MC grappled to catch him and spin him toward the wheelchair, which was improperly locked and rolled away a few feet. Somehow, MC managed to land Festus perfectly in the chair's seat. As MC looked up, he could see from the expressions on the faces of a small group of people that it had been quite a landing.

Festus seemed to no longer be participating in a physical way and had an unconcerned look on his face.

As MC and Festus were waiting for the admittance paperwork to be completed, Festus complained of a pain in his ribs.

MC and Festus passed through the halls lined with paintings. It was a gentle and relaxed ride to the nurse's station.

At the station, the desk nurse looked at Festus with a scowl and then looked at MC and said, "He doesn't look good."

MC quickly said, "He's not been doing so well."
He thought the nurse was questioning whether Festus should be admitted. He wondered how a hospital could turn someone down because they didn't look good.

Then the nurse exclaimed, "I mean he doesn't look too good!"

She pushed a button and people in scrubs came scurrying in all directions. They lifted Festus from the wheelchair and carried him toward a room.

"Zero pulse in left arm!" someone yelled.

In the commotion, MC realized what had happened and whispered to Festus, "You've gone."

The veterans' hospital took charge of things. Since Festus had been admitted into the building before death, they would take over and help with the cremation and all the necessary paper work. No one there could remember such a timely departure.

Gladice's name was listed as Festus's next of kin. MC's power of attorney ended at his father's death. It was hospital policy that the doctor calls the next of kin with news of a death.

After an hour and a half of calling, the hospital could not locate Gladice, who, it turns out, had gone on a trip.

Finally, MC was approached by an administrator, "MC, if you can assert that you have cared for Festus for at least six months, you can sign these papers."

MC bade his farewell to his father and signed off. Festus's body was draped with a flag and placed at one of the entrances of the hospital to be given tribute. Festus had gone with no struggle in a very precise and timely fashion.

MC spoke at Festus's funeral where many Macs had collected. He was limping with gout in the knee due to his chocolate consumption. He was grateful for the lessons that Festus had bestowed on him. He had no regrets. He could now phase out the conflicts with the Grabbits. Though he made no deliberate effort to avoid anyone, MC had asked inwardly to not have to deal with any of the Grabbits on that day. As it turned out, they would not even make eye contact and came nowhere near MC. The reception was a Mac family reunion.

Toward the end, MC's cousin Frick made his way toward him. Frick was not a social talker but somehow he got started. "The Grabbit boys went to see someone and they said you don't stand a chance."
"A chance for what, Frick?"
Frick went on, "Gladice is a little short on cash and needs a place to stay."

MC did not want to get angry with Frick. As if from a distance, he heard himself gently replying,

"Gladice left the situation and now she wants to come back?"

Frick said, "She was sort of intimidated."

MC did not mention the letter from Missle Toes, but said, "There was a loan taken out to repair the homestead and since she has life tenancy, she would be responsible for paying it back if she lived there, plus the interest plus the electricity, phone and insurance. Life estate doesn't mean I would have to pay for her groceries, you know."

Frick groaned as he looked at the floor. Frick had expected something different, maybe an argument from MC. Instead, he was given a maze of information to take back to the Grabbits. It was going to cost something. Frick was just the messenger and he wanted out of the conversation.

MC closed it by saying: "The days of Festus giving everything away are over, Frick!"

The Grabbits had inquired how they might get compensation after Gladice had left the homestead. They would not come forward as long as Festus was alive.

While the Grabbits were huddled only twenty feet away in a corner of the church basement, they had chosen Festus's funeral and picked MC's cousin as their messenger boy in their eagerness to catch up on a year and a half of missed opportunities.

Some of the Macs were awed upon hearing of the compensatory claims. MC's brother Toulouse took this news rather hard and looked exhausted. He assumed that things might clear up after Festus's death.

On a few occasions, Toulouse would ask MC if he had received a registered letter in the mail yet. Within a month, Toulouse had barely managed to get down to MC's and collapse, only to end up at the Hill Top hospital.

Toulouse had concealed from almost everyone that he had been having headaches and vision problems. Having no insurance, he had toughed it out too long. There was nothing that could be done about a tumor lodged deep in between the lobes of his brain. MC would witness Toulouse having his fatal grand mal seizure. The seizure was caused by metastatic renal cell carcinoma to the brain.

Toulouse told MC just before his seizure, "I'm taking over where Festus left off."

It all happened quickly. Soon MC would make the decision to have Toulouse's respirator turned off.

Just one month after Festus's death, the Macs once again gathered for a funeral. During the ceremony, Frick and Malice went to the homestead and took more of 'their' things out of the barn.

A month or so after Festus's and Toulouse's deaths, Festus's car, which had Gladice's name on the paperwork, vanished from the homestead driveway in plain daylight, leaving MC with no way home. When MC went outside, the wind was blowing snow into the fresh tire tracks. A note had been written on a brown paper bag that was held down with a piece of scrap lumber.

"MC, the car is being returned to its rightful owner. Please refer to the enclosed documents. Your personal property will be returned to you. The state police have been informed of this repossession."

Whenever the car had needed repairs, insurance and registration, Malice would declare, "We don't want the car." MC had come to expect her spiteful reactions. Fortunately, the phone was still working and MC could call out and get a ride away from the homestead. MC had been paying the phone bill for the past two years. Soon the phone stopped working and while at the homestead, he could not make any calls. MC convinced the phone company to temporarily restore service until they were sure who was paying what. MC warned the phone company personnel that Malice was difficult. They called back and affirmed that she was very difficult, but that the current account needed to be disconnected and a new one reconnected. The phone company did not carry out this transfer fast enough. In a fit of rage after MC had gone, Malice came to the homestead and ripped the phone wire off the house.

The Grabbits were gone. MC eventually heard more stories, each more astonishing than the next. A small community seems to ferment over such things.

MC heard that Frick and Malice had a successful auction of their farm equipment and had built a large new house. Now they were destitute and both had to work full time. They blamed MC partly for their supposed hardship because MC had taken their mother to the cleaners. Gladice would tell of the tears she shed because she and Festus were not allowed to celebrate their anniversary. Finally, she lamented on how MC had withheld the news of Festus's death from her.

The Mac homestead had been spared on several occasions and the Grabbits themselves inadvertently created a situation that saved the homestead. By holding faulty and secretive deeds, they would actually protect the homestead from being sold and from further partitioning or subdivision.

The Grabbits made it necessary for the creation of a protective trust. Otherwise, Festus, stripped of his goods along with his destined path to the rehabilitation center, would have created liens on the homestead. Toulouse's uninsured stay at Hill Top Hospital would have taken the rest of the homestead along with him had it not been protected. Had the Grabbits left things alone, the fate of the homestead would have been worse.

MC's ancestral link was far more than just the homestead. His father had exemplified all the hardships of the area. Most of the Mac's had left the area of the homestead. Its disrepair was symbolic. When MC took on Festus's care, he realized he was dealing with the human condition of several generations. It was an attempt by MC to "Make your blood relation's heart glad."

Malice would later make headlines. She was concerned about the potential variances in the town budget. Malice, as if campaigning, accusingly stated that there were discrepancies in the town budget. People were a bit puzzled since Malice, who had been auditor of the town for the past twenty years, was now questioning accounting practices.

It was during these same years that she had been in charge of Festus's accounts. Reading from a prepared statement at a town meeting, Malice would describe the role of an auditor and shrug off any blame. Nevertheless, Malice, with her cheeks bright red,

announced that she would be stepping down as auditor and would refuse any farewell gifts.
The homestead is located in a depressed, downtrodden area where economic prosperity is slow in coming and where economic hardship lingers, often resulting in dysfunctional behavior among the area's population.

Neighbors came to visit MC and claimed that Festus had something of theirs or that he had promised them certain items. Most would declare that they were watching over the homestead, only to be later involved in the pillaging of scrap metal and old trucks.

MC's good friend Bess would say of the Grabbits or the neighboring thieves, "Don't take it personally, MC, they would do this to anybody."

MC could not imagine a more difficult and stressful year to challenge the effectiveness of this cancer treatment. However, the cancer numbers of Dr. Heil Mittel Mandel's hair analysis on MC continued to improve. Dr. Mandel had increased MC's dose of the pork pancreatic enzymes. The doctor noted that the only lack of improvement was in the stress organs and nothing to do with cancer.

The same period included the passing on of Popsy who had helped MC find information on alternative cures for cancer, but then passed away quickly following orthodox cancer treatments.

Also during that time was the passing of Sharinagar, who had helped MC to perceive relationships differently and remain determined to see things differently.
Along with his father and brother's deaths, these four important figures in MC's life all passed away within a

few months of each other. Two were in their eighties and two died of cancer in their sixties

MC had been involved in the care of all four.

Due to a resolve that was established in MC shortly after his diagnosis of terminal cancer, he managed all of this while following a diet and a routine that were imperative to his continued good health. He took several thousand enzymes and vitamin pills per year and ate a mostly raw and organic diet. He followed a schedule of purges and cleanses and enemas all while struggling with financial insecurities, family dynamics and death.

That year, MC would often think back to the origins of this resolve. While lying in a hospital bed, he had attained this state of mind, not by trying to change his thoughts, but by letting go and asking himself, what is the lesson in this? What part did he play? What denials were being reflected back to him?

MC began to realize that stress was a mental interpretation. If anything, his avoidance and reclusion created more stress and rage. The decision to do the right action would calm his stress and would affirm itself by having events fall into place.

It also affirmed that special personal adage offered to MC by the late Sharinagar: "Where there is trust, there is faith."

"A challenge strengthens the mind. Problems weaken the brain.

"All you have to do is be true to one thing in your life and it will bring order to everything else. It will transform you."

Tara Singh

13
ENERGY

MC felt that there were turning points in his life. Being diagnosed with elevated stages of colon cancer was a definite turning point. It could eventually be seen as the best thing that ever happened to MC.

MC was always one to think that there were no coincidences in life and that all things he encountered would have meaning…someday. Perhaps all humankind was headed in a common direction but each had to experience it differently. MC's discovery of life's energy was just one such turning point - something that would be needed as a backdrop to MC's survival.

It was by now the early '70's. There was a lot of political terminology that slowly gave way to psychological terminology. The term "energy" was just beginning to be used in vague and mystical ways. MC was perhaps fortunate not to have had a vigorous religious background or a dogmatic political upbringing. This gave him more liberty to inquire without guilt or prejudice.

On the day of a large scheduled student demonstration, MC found himself wandering and finally sitting on the floor in the back corner of a book store reading for hours. He had serendipitously stumbled on the selected writings of scientist Wilhelm Reich.

Interest in Reich had been amplified by his imprisonment and death shortly before his supposed release in 1957. The 1960 FDA-supervised burning of his publications on "energy" drew further attention. It was as if information about the "orgone energy" was forbidden.

Basically, Reich stated that humankind lived in an ocean of background energy that swirled, pulsated and flowed through them and all other matter from the tiniest plant to the largest galaxy. The tremendous energy that MC acquired by simply reading these works, seemed to make it evident that this energy could be amplified and released by discoveries or breakthroughs in the mind. Sharinagar would often later say that certain truths would silence the mind.

MC could read a small paragraph of Reich's writings and then ponder on his tranquility, unaware of where he was. MC would buy all Reich's books and begin a new flight into the "energy" phenomenon. It, too, would become a wave. By the '80's, MC found himself attending energy fairs.

The Chinese called this energy "chi;" in India it was "prana," in Hawaii "mana." Germany's Karl Reichenbeck christened it "od," Franz Mesmer called it "animal magnetism" while Reich called it "orgone." To Rudolf Steiner, (of the Anthroposophic movement), it was "chemical ether" and Russia's Inyuhin called it "Bioplasmas."

> Sharinagar often talked about this energy as "prana." This was an ancient term and a bit more digestible. He warned of the distortion by folks who knew nothing about it but who would mint money in the name of it.

Energy's abundance and mysteries opened up all sorts of possibilities. The most astonishing phenomenon was the great fear of this energy. Was this perhaps because it required the use of a dormant sense?

This esoteric knowledge was best accepted when veiled in non-scientific folklore or fictitious-sounding novels. Reich, of course, made the mistake of being analytical. His discovery of orgone came to him through his psychoanalytic practice. He observed that whenever a patient let go of one of these character armors, (unconscious tension that is locked in the patient's emotions), there was a massive flood of emotional energy. He also made the mistake of delving into many areas of its applications. The area that possibly did him in was his application of orgone energy to cancer research. He had noticed that during the emotional release while undergoing character analyses, some patients would discard existing cancerous tumors. Reich felt that this energy could be accumulated, drawn off and redirected to create, through self-awareness, emotional situations that could help in the acquisition of health.

This cosmic energy theory seemed to bring about negative reactionary outbursts among the armored and fundamentally neurotic disciplines of the time. Reich would call this the "emotional plague."

It was at the same time, (1936), that a Russian named Semyon Kirlian would discover this electrical aura surrounding the body. Kirlian made photographic plates of this energy around the hand or around a leaf. What intrigued MC about Kirlian photography was the existence of phantom images. When a leaf was photographed with a missing section, there were sometimes manifestations of energy observed on the plate in the area of the missing section. Reich would categorize the universal life force as having various concentrations depending on whether it was a living organism or mineral matter.

Reich described this energy as manifesting itself within nature in numerous ways such as the flow of the river, wind currents or blood arteries. It dramatically manifests in helixical and torus-like shapes such as tornadoes, hurricanes, galaxies as well as spiraling vines. This often fertile energy will solidify itself in solid, superimposed shapes and forms resembling the fetus or kidney-shaped organs and seeds found throughout the living world.

He described this energy as even flowing through the mental, creative and psychic thought system of man.
The perception of the energy had made its way into Impressionist paintings such as Vincent Van Gogh's "The Starry Night" or the spotty textures of George Seurat's Pointillism paintings.

The existence of this energy would begin to give MC some insight into the rituals and symbols of organized religions all over the world.

In his early discoveries, Reich was astonished that a critic would mock his findings and boast that "Billy" Reich would next proclaim the cure for cancer. Reich would never proclaim a definite cure for cancer. He referred to cancer as a riddle to be solved. He was always amazed at the prophetic criticism that would precede him to his downfall.

At the time, MC was either too timid or unable to put this theory into play. Instead, since it was inferred that applying such theories without being properly licensed was illegal, MC would study science and pre-medicine. After all, the Federal Drug Administration (FDA) had put

Reich in jail for crossing interstate lines with his orgone box accumulators.

Orgone was Reich's derived name for the bluish-tinted energy that surrounds the planet. He explained that the energy is drawn in by organic matters such as cotton, wool, hay and wood and is reflected by metals such as iron, copper, silver and gold.

By constructing a box with alternating layers of cotton and metallic materials and placing sheet metal on the inner side of the box and a wood casing on the outside, a slight accumulation of this pulsing energy was directed toward the inside of the box. A box large enough to hold a person could then be beneficial and used by patients during Reich's psychoanalytic therapy. This would evolve into Wilhelm Reich's "orgonomy".

The description of this energy did seem to bring up controversy, particularly from those who already had set ideas about how atoms and galaxies work in a flat, two-dimensional view.

It did not help Reich's credibility to be exploring other fields such as weather control and extraterrestrials. He felt that extraterrestrials had mastered the use of this energy as a propellant. Reich would devise a small, motor-like tube that ran spontaneously off this life energy.

His earlier work on sexuality, and his interpretation of sexuality as depending on the build up and release of this primal energy, gained him notoriety that would last from the Nazi period in Germany all the way to the Eisenhower era, during which he would die in prison.

MC, in those early days, could only suppose that Reich had made the big mistake of being too open and scientific. The scientific industrial class takes offense to new uncharted discoveries, especially since the energy would be free for the taking. MC would recognize a similar defensiveness among the oncologists when discussing alternative cancer treatments.

MC remembers at the onset looking through an iron pipe on a very dark and clear night. He saw a corona light that resembled a bluish propane flame twisting and sparkling all over the inner surface of the iron pipe. MC tried to show some of his friends the phenomenon and soon noticed a resistance toward it. Some did not even dare to look through an iron pipe. Others explained it away as an optical illusion caused inside the eye. But one of MC's good friends, when looking at the pipe said, "I remember seeing this when I was young. I saw it a lot and mentioned it to my doctor. He gave me some medication that dulled it a little bit. I finally told the doctor that I didn't see it anymore."

MC was once asked by a science buff to give examples of manifestations of this energy. MC recollected the occurrence of these energies during those dark nights, whether as lightning bugs, northern lights, St. Elmo's fire or glowworms. One night, after a rain, a rotten log was glowing an intense blue light which MC found out later was called "fox fire." Often times, right after a snowstorm, MC would poke his arm into a snow bank and observe, even on a gray day, a blue aura inside the tunneled out hole.

One evening, MC went swimming in the ocean and was amazed at a glowing in the water. When he dove under, the bubbles triggered a bluish wake. When his hand touched a round soft jellyfish, it would light up into a blue sphere under the water. The ocean water would even sparkle as MC poured water down his body. It was surprising that so few people knew of this and those who did simply define it as "phosphorescence." MC then described seeing small spiraling white spots streaking around in his path of vision when gazing at a bright or clear blue sky. The science buff replied dismissively, "Those are just dust particles stuck in the mucus layer of the eye."

MC remembers Reich stating that when looking through a microscope at a magnification of 2500X, and preferably above, the pulsing turbulence of the organism became so magnified that it became difficult to get good resolution. Reich felt that much could be diagnosed by observing the vibrancy of cells under high magnification. The distortion in these high magnifications frustrated researchers and would lead to the creation of the electronic microscope, which froze everything still in its tracks at very high magnification resulting in the resemblance to a lunar landscape. This is the direction that the science textbooks were taking as MC went back to study the sciences.

What the Curies discovered in their work with radium stole the show. This was something that could be mined, enriched, isolated, concentrated, and used for such things as x-rays. Once more, it was black and white and motionless, something concrete that science could manipulate. Its radiation was even foreseen as a cure for cancer. Even though Madame Curie would die of cancer caused by her exposure to this radiation, the idea persisted onward to burn away cancer with the substance that could cause cancer. Due to its heat-producing reaction, it became a possible fuel. Even though Albert Einstein stated that nuclear power could never be used in a practical way, it is being used anyway. Even though there is a strong reaction between, let us say uranium and this life energy that causes heat, the existence of the latter is still contested, doubted or just plain denied.

Partial descriptions of this mysterious energy have been written of by authors who have veiled their works in fantasy novels in order to fend off the scrutiny of empirical science - a science that sees this spiritual science as the superstitions of ancient, primitive cultures; the same way fundamentalist religions pontificate on the use of paranormal forces in old scriptures, while frowning at the use of the godlike energy in today's world. The fundamentalists are incredulous at its application by the common man, speaking of it in gospels as a phenomenon that could only have existed then but not now.

The ethereal energy that is everywhere cannot be grasped easily. It stands little chance of being understood by the untrained senses of empirical sciences. Their current mindset is very precise in counting the carbon particles trapped in the atmosphere but have little understanding of what radioactivity and its reaction with life energy will do to change weather, health, seismic activity, the creation of deserts or to enhance climate change itself.

Nuclear reaction has been given the uncontestable label of low carbon, clean and safe energy.

Not so long after his discovery of cosmic energy, MC found himself in the middle of a quandary. He had great fervor for this bio-energy and all of its possibilities. Yet his courses in psychology and medical science were always limiting the expressions and wanting to sedate the symptoms. The processes of body elimination were suppressed. The fervent body chemistry of this bio-energy was repressed by blocking and not allowing expression, especially in dealing with mental illness or emotionally-induced outbreaks. Somehow, the information about bio-energetic science to help humanity was only available on the shelves of occult bookstores.

MC could not reconcile the two thought systems. His personal inclination about this cosmic energy he perceived when young eventually went on hold. Though he still participated in conferences to hear about healing energy, earth energies, spiritual energy, psychic and channeled energy, it was more like a book club review than a passion.

One day MC would awaken in the emergency room with a colostomy and category-four colon cancer. MC would find himself at the mercy of a medical science that gave him very little chance to survive this predicament and zero chance to cure it. Having only in mind what he had discovered during his energy quest, MC had to formulate a decision. It was apparent that the application of energy principles needed to be practiced.

"Don't talk of these things until they are your direct experience," Sharinagar had told MC. He deplored those who read about energy or kundalini and then open up shop to appear as experts.

It was an internal decision as well as a lonely thought system. Most of the world denies any cancer cure as well as the existence of a creative cosmic energy. Others mystify or fear it believing that to play with this energy is risky.

An important contribution of Reich's work was his associating the unconscious to such diseases as cancer. He felt that as soon as there was a blockage due to the body armor, this orgone energy would dam up in that particular area. Blood, plasma, oxygen and energy were all involved as one. When this area was starved and perhaps even inflamed, it would become prone to such things as cancer.

Reich broke down the body expressions into two tendencies: contractive or expansive, with contractive being more constrictive (phlegmatic, arthritic), and in need of more energy and expansive being more explosive (high blood pressure, hyperactivity), in need of less energy. Reich's seven layers of body armor, which were vaguely correlated to the "chakras", had to be free-flowing for a healthy body and mind. These armors, according to Reich, seemed to be induced by childhood traumas or sexual repression. One of these armors, the fear reflex of holding one's breath, could create vulnerable areas in the body for degenerative diseases that thrive in the stagnancy of low energy and low oxygen.

Wilhelm Reich practiced psychoanalysis. He succeeded in dissolving these armors in certain patients. MC had met someone whose aunt's cancer was dissolved when going through this therapy. Successful cancer treatments accomplished without purges and cleanses could lead to clogged kidneys and Reich was well aware of this fact. Reich was never terribly encouraged by his results using character analysis since so few breakthroughs could be accomplished in a short time and patients resisted such changes in their character armors. Sharinagar once quoted to MC words from Jiddu Krishnamurti, "People seldom ever change."

Both Reich and Krishnamurti saw hope stemming from the education of children that would avoid these psychological armored blocks. The direction to be taken by Reich, as well as others, was to create energy-directing devices or apparatus as well as nutritional education to enhance the effects of such treatments, despite psychological resistance.

MC found himself in his predicament despite his knowing of these theories. All his studies of social, political, psychological, psychedelic, sexual and cosmic energy did not spare him from cancer. MC was now searching for "a cancer treatment," a treatment that would not include cancer-causing agents and that would include all manifestations of life energy that he could discover.

MC was not Internet savvy and the information he needed seemed to unfold as he was ready for it. His fatal situation helped him to become determined. MC had been alerted by Reich's work that ionization, caused by television, fluorescent lighting, transformers, microwave ovens and electric power lines, interfered with this energy. The energy's name was referred to by now as the "body electric".

Science could only relate to this energy by its interaction with electric devices. Kirlian photography, for example, could only be done by the intrusion of electricity. Photoplates done without electricity were far less dramatic. Was the aggravation of this life energy needed to visualize its existence? Reich used the term "confounding" when explaining the relationship between cosmic energy and other wavelengths coming from electric or radioactive devices. The agitation, contamination and perhaps fission seem to create the perfect environment for increasing the occurrence of cancer.

While MC was beginning to gain strength, he came across the discoveries of embryologist

Dr. John Beard. Beard discovered small packs of embryonic germ cells, distributed throughout the body, whose sole purpose was to activate their stem cells for the repair of the body. MC now imagined how the intent of these cells exposed to electromagnetic pollution sensed panic in the body. These confounding electromagnetic anomalies would stimulate and distort the messaging to the embryonic code producing trophoblasts that would lodge themselves into the most vulnerable inflamed area of the body. The trophoblast, while mimicking certain reproductive principles, along with its misguided stem cells, would end up giving birth to a monstrous blockage of MC's intestines.

The first thing upon awakening from surgery, MC found himself having to learn how to breathe. MC would have to deal with the reflex that he had read about in Reich's work. The body, fearful and anxious, tends to suppress its breathing. Adding to this, the forgetfulness or the unwillingness to drink water during this turmoil left MC deprived and exhausted.

The Hawaiian healer, Kamili, would later tell MC how he should breathe: "Inhale for seven seconds; hold for seven; exhale for seven and hold again for seven seconds."

She also told MC that to bring him back to the divine light he needed to consume papaya every day.

Right after surgery he was asked to blow into a plastic device called an incentive spirometer. It had a little plastic ball that went up and down, measuring the strength of his breath. Since MC's body had adjusted to a shallow type of breathing, blowing into this device was a burden.

Later Sharinagar would show MC the basics of pranayama. Sharinagar warned that the practice had been commercialized as a preliminary to exercising. Its function was, according to him, to accumulate the prana in order to give oneself the strength to enter into awareness. To be peaceful, still and quiet requires a certain amount of energy, which then regains an abundance of that energy for one's well being. MC would experience the capability of the brain in directing the energy.

What Sharinagar did during his sharings was to bring people to this particular state where there were no thoughts in the mind. MC felt elated by these sessions. It followed the laws of this energy in which both the giver and the receiver were energized.

But MC was puzzled by the lack of any obvious technique for this transfer of energy. He even purchased the cassette tape of the session hoping to relive it. The presentation of truths as facts seemed to trigger this thought-free state that guided one to be stress free.
The subject matter seemed to have no bearing on it, and replaying the tape did not necessarily bring on this grace. Also, the state seemed to fade upon returning to his routine life. To MC it seemed to require a certain repetition and commitment to this burgeoning state. He saw it as a form of mantra-like existence and found himself remembering the experience but no longer able to re-experience it at will.

MC noted that while his spirit longed for this state, his will lacked the energy to aspire through his senses. Forgetfulness, distraction, denials, guilt and compelling appetites preoccupied his senses, postponing any inkling of awareness. MC felt that this loss of willpower was the status quo behind the quagmire of organized religions. Sharinagar claimed that this altered state came about due to what he called "undoing;" breaking down one's judgments, one's prejudices and questioning one's beliefs in order to be receptive to the laws of this rejuvenating energy.

It occurred to MC that the medical world would be limited in their application of medicine without knowing the mysteries and miraculous functioning of the lungs in capturing the inspirational energy. Instead, the preoccupation was on containment of symptoms and on the germ theory. French chemist and biologist Antoine Bechamp stated in his work ("La theorie du microzyma, 1888") that the airborne germ theory has hobbled the medical science in secular error and has become an obstacle to the understanding of the organization of life and its environmental influences.

MC would sigh when thinking about Lollypop telling him she could eat no raw food during her chemotherapy, take no vitamins or undergo any cleanses or enemas. Since the chemotherapy overloaded the immune system, all foods needed to be sterilized to avoid infections. As it turned out, all eating and drinking became nearly impossible for her anyway. MC felt fortunate to be following a protocol encouraged by Dr. Mandel. When MC found himself dealing with such conflicting theories, the doctor would remind him that "the benefits of raw foods far outweigh the detriments."

This suffocation of the flow of energy most likely gave birth to the origin of MC's cancer. It would then cannibalize his body for its energy and growth. In his early fifties and condemned to colon cancer, he had to find a way to survive by consuming the cancer -- similar to the way the fetus consumes the placenta for its survival.

While waiting for his first visit with Dr. Heil Mittel Mandel, MC established a routine. He got the old orgone box that he had made when in his twenties and placed it in a barn open to the outside so as not to contaminate it with indoor appliances and such. He placed bales of hay all around it and on top of it. It seemed to MC that it pulsed at times, especially on a clear, anti-cyclonic crisp winter evening when the stars twinkled. Upon the approval of Dr. Beauve, MC practiced his breathing exercises while inside this box. MC was fortunate to live in the hills of Vermont.

Unaware of Dr. Mandel's program, MC was unknowingly already consuming some juices, including four pounds of wheat grass a week.

When MC went for his first visit with Dr. Mandel, the emphasis would be on nutrition. Dr. Mandel nodded and smiled when MC told him of his breathing routine in the orgone box. It was all that MC knew of at this point. MC's type of cancer needed an alkaline diet, which would require 70% of it to be raw. The doctor affirmed that wheat grass juice is the most alkaline food a human can consume. Most of the 70% raw diet was realized by consuming a quart of raw (mostly carrot) vegetable juice per day. After twenty minutes, the vitality of the vegetables begins to disappear so it was vital to consume it within twenty minutes of its pressing. What could this vitality be but a live energetic charge?

The processing of these raw substances will often create a product with a golden amber color, such as in honey, apple cider or maple syrup. These substances will retain some vitamins and a caloric value. The energy that helped fabricate these will vanish, especially after fermentation or sterilization. For the most part, it tends to acidify the food. Most of the time this energy of fabrication vanishes back into the ocean of this creative cosmic energy. The only experience that MC had to compare to the alkaline diet, its purges and cleanses was when he had fasted years before. Drinking only water during the fast put his body into a cleanout mode. Being deprived of any foods put his body into the mode of processing and discarding excess, built up fluid, fats, salts, minerals and, presumably, toxins. Despite the lack of food, MC found himself alert and, at times, very energetic. During these fasts, MC would spontaneously start cleaning off his desk or organizing his bedroom. The deep sleep during these fasts were similar to that experienced during Dr. Mandel's moderate vegetarian, alkaline diet as well as his liver cleanses or citrus purges. Among the miraculous results from fasting noticed by MC was clean skin, clear eyes, good blood pressure and a trim look. This was all quickly erased by binging on unfavorable food.

MC's compliance to Dr. Mandel's protocol diet was essential for accumulating the energy needed to keep his cancer-free state. For the first two years of the cancer treatment, MC was a purist about his prescribed diet. He was alert, slept deeply and was observing his body's response to discarding the cancerous state he was in.

MC would modify his previous history of extremes of going from a water fast to a complete, gluttonous feast.

More could be accomplished by following a selective, strict alkaline diet, specific to MC's individual needs. This would balance the energy to his autonomic nervous system and accomplish the detoxification that occurs during a fast.

MC's mostly alkaline raw diet guided his energy toward his parasympathetic functions: digestion, the circulation of blood flowing toward the digestive organs and the improvement of skin circulation. The most notable was the increased secretion of enzymes. As a result, MC would awake in a very drowsy state. It was not like him at all. Upon calling Dr. Mandel, MC was informed that this was a very good thing for his type and temperament. His energy was being diverted to where it had been neglected.

MC had always been on the go. He used to wake up alert and would fire up into a multitasked life style. An acid diet helped maintain this almost manic existence. Dr. Mandel reassured him that this deep sleep state was favorable for his tissue repair and his purging of toxins. The alkaline state was unfavorable to his type of cancer, which thrived on acidity. In an attempt to find equilibrium between the autonomic temperament of the body, certain foods were suggested.

Enzymes were what MC needed to take in abundance. Interestingly enough, papaya is loaded with digestive enzymes. Was it not these enzymes that were the link between the body and the energy inside the food? If the food and enzymes he decided to eat directed where the energy went, wouldn't it be important to consume live, energetic foods? The inclusion of their enzymes would give the mind enough energy to regulate existing energy that enveloped him as well.

MC remembered, upon awakening from his second surgery, a feeling of being trapped inside his body. He assumed it might be side effects from the medications or maybe impatience at getting started on his new enzyme therapy. He had never felt it before and, as it persisted, MC realized that, with low energy, one is in fact a prisoner inside the body. When his body acquired more energy and health, he was uplifted and freed from this feeling of encasement.

It was others who brought MC's glowing appearance to his attention. The healthier he became, the less preoccupied he was with his body. MC was very lean. He had acquired a bronze tan, which was a combination of carotene from the carrot juice and the sun. He had become rather pensive and stared at people with an intense, dynamic gaze. His hair and beard were growing stronger and darker. Since the word was out that MC was supposed to be dying of cancer, it made people take notice even more. This inquiry from many gave MC the urge to share. He found that those who boasted prematurely about their healings were thought of as mythomaniacs. MC thought of them as unhealed healers. He had been guilty of this himself. Now MC, who really had been rejuvenated, explained this in an especially grounded manner, and shared only when asked to do so. This was the result of Dr. Mandel's intuitive nature, which amazed MC. The doctor always used language familiar to MC's thought system. MC as a patient needed to formulate the proper question in order to get Dr. Mandel to dole out more insight. This turned out to be a self-protective measure for the doctor. More metaphysical explanation would surface as MC developed a relationship of trust with the doctor.

Dr. Mandel did not recommend any particular mystical practices. He felt that someone following a proper protocol would find his own understanding and expression of this compassionate source of energy. The greatest insight that Dr. Mandel gave to MC was to point out that it was MC's decision to be compliant that cured him. Had MC fallen by the wayside because of fear, and had he never experienced the sensation of this energy reclaiming his rightful body, he would have died not knowing the endless possibilities of this metamorphic energy.

It was suggested that he should have chemotherapy. However, even this suggestion was dropped for MC's case, as Dr. Navaja somberly suggested to MC, "Go take a cruise!"

In a way, this made MC's decision easier. It was only when MC was declared helplessly afflicted that his state was brought to his true awareness. The health of his body had to be an internal decision and his internal voice would guide him to health.

MC would come to feel, whether in treating cancer, dealing with agricultural problems, social cohabitation or finding a renewable source of energy, things always circled back to this mysterious life force energy. To be able to harness this energy would seem to be the only way out of the world's interwoven dilemmas.

While following Dr. Mandel's diet and purges, MC became interested in how all these purges worked. Thus, Dr. Mandel's literature became MC's reading material. Breakthrough after breakthrough, MC began to witness that the purges were enhancing his health. Even the skin brushing, which seemed more like what a mother cat would do to her babies, seemed to be effective. This was one detoxification routine that MC avoided trying to explain. To some people, the enzyme therapy seemed like hogwash as it was. Brushing one's dry skin with a loofah sponge up each limb and down the neck and trunk toward the lower abdomen was indeed "energy work." This procedure was more vital toward the beginning of his therapy when the lymph nodes and skin seemed loaded with toxins.

Even though MC was not interested in the politics about energy while dealing with his cancer, he could not help but notice the hostility toward it. From Franz Anton Mesmer to Nicola Tesla, all the way to Wilhelm Reich, there was a calculated effort to disprove the existence of their energies. When these efforts to persuade the public's belief failed, political, private, and/or government agencies were used to crush the information. As MC's cancer treatment progressed, he would stumble onto more of these victimized pioneers. This exploration became the "cruise" that Dr. Navaja had suggested after reading his dismal CT scan report.

MC was amazed at the results of research scientist S. Meryl Rose's experiment done in the 1940's. Rose worked with salamanders, which have the ability to regenerate amputated tails or limbs. Some form of energy current converging at the blastema, (a mass of cells at the stump), regenerates itself. If this current was repressed, there was no regeneration. It was discovered that when salamanders, having had a tumor transferred to their limbs, were amputated through the cancerous part, something amazing occurred. Not only did the salamander live, unlike the malignant un-amputated control salamanders, but as the amputated limb started its regeneration, these salamanders were cured of cancer. Upon closer observation, it was noted that these cancer cells were returned to normal by dedifferentiation and were imbedded inside the new regenerated limb. Considering that the tumor cells had come from a frog, another species, MC had the impression that regeneration and perhaps a cancer cure were more dependent on an energetic manifestation than on a cell's DNA code. The differentiation of cells in guiding the regeneration of salamanders' limbs all the way to the end of their finger tips, would seem to indicate the presence of an energetic ghost field.

Now Kirlian's photography of the phantom limbs did not seem so bizarre to MC. According to Wilhelm

Reich, cancer would tend to occur where there were energy blockages, either physical or psychological. MC wondered if it was the energy's quality and quantity that affected signals to the DNA itself.

Dr. Stanislaw Burzynski discovered that certain peptides, (antineoplastons), were missing in cancer patients' blood and urine. Thus, the energy flow could not transmit signals to inhibit DNA synthesis and cell division in cancer cells. This led Dr. Burzynski to call cancer, "a disease of information processing."

MC had been told that colon cancer would often come back to the scar tissue where the tumor had been removed and where the two severed ends are reconnected. This made sense to MC, since a thick scar could stop a normal energy flow. Enzyme therapy seemed to have compensated for this surgical scar.

Dr. Mandel told MC that the regeneration of bone had been accomplished and enhanced with the use of a device that concentrates electromagnetic energy to the end of the amputated limb. In man's case, this energy needs to be concentrated and directed toward the end of the amputated limb. Dr. Mandel went on to say that this would require leaving the end of the limb somewhat open; something that would go against all formal medical training of cauterizing, searing, tying off and suturing the end of the amputated limb.

MC remembered twenty or so years ago, going to a dowsing conference and hearing Christopher Bird speaking about his books, "Secrets of the Soil" and "The Secret Life of Plants." Bird started off by proclaiming that Gaston Naessens was being persecuted for his work, which was similar to the experiences of American inventor Royal Rife. Bird shouted out, "What became of the Rife microscope?" These conferences usually had so much material and created such information overload that MC could not process much of it for long periods. Though MC remembers Wilhelm Reich stressing the importance of high magnification to observe blood, Bird's question did not immediately inspire MC.

The next time MC would hear about Rife was during his cancer affliction. A friend would tell MC that Royal Raymond Rife had invented the Rife Frequency machine. He could adjust it until it dismantled cancer in some way without affecting the healthy body cells. His friend's information was a bit sketchy.

There seem to be many coexisting scientists in the twentieth century approaching this cancer problem at different angles.

Rife's story was an epic so far ahead of its time that it appeared to be science fiction. He had invented a light microscope that could observe living microbes at a power of 17,000 times. The highest at that time was 2,500 times. Rife found what he called a guilty microbe, which always accompanied cancerous tumors. He would call this microbe Bacillus X or BX virus. Wilhelm Reich called it "T-Bacillus." Rife claimed to have seen its mutation on his microscope, a behavior named pleomorphic microorganism theory or "shifting form". Rife asked himself if this live organism, producing a purplish-red color, and therefore having a frequency, could be killed by another frequency, which would resonate with the vibratory rate of the microbe. Rife created a ray tube, which broadcast various frequencies. When it was finally ready, he would sit in front of his microscope tuned to the BX microbe for hours on end, tuning the dial of his frequency device, going through one frequency after another. One day when he reached a certain frequency, he saw the light of the BX microbe grow brighter and then go out, after which it disintegrated. Rife would call this the Mortal Oscillatory Rate (MOR) for the cancer microbe. This did not fall into the germ theory way of thinking. It was not an airborne germ. The Bacillus X did not cause cancer. It came about as a symptom of cancer. The BX's destruction, by means of a frequency of 434 hertz, was accomplished by intensifying this frequency in the energetic field of the patient. The elimination of the BX was done by the exposure to this energetic field, as if the patient had lost the strength or ability to generate this frequency. This method was successful in dissolving cancer in humans.

As MC's friend had told him, there were persecutions. Rife's laboratory was set on fire, burning film explaining Rife's work in detail. There are two of his microscopes in museums, but the essential parts are missing.

As MC's friend summed up his interpretation, "They weren't able to confiscate all his equipment, so the technology is still out there."

This ray tube broadcasted frequencies through a tube filled with helium or argon. It was referred to by a child as, "the blue light". MC was quick to note the same color as what Reich called orgone. The tube never made direct contact with the body. It was applied from a few feet away.

MC pondered on his brother Toulouse and his mother Lucette, both of whom had brain tumors that were inoperable and untreatable by standard medicine. What harm could this have done in his mother and brother's cases? Especially in his mother's case, who was sacrificed as a medical exercise with no probability of ever living a normal life.

The well-funded industry of standard cancer research continues to seek patents and cures all in the name of protecting the public.

How many millions could be helped if some of this funding was directed toward the research of less standardized cures rather than wholly aimed at the monopolizing field of standard medicine?

MC remembers being told by Dr. Navaja that chemotherapy and radiation were the "only proven effective treatments against cancer." MC was also told that in his case it would only be palliative care, but that he should avoid wandering around looking for alternative treatments and their "herbs."

In the end, it seemed to MC that true health would be determined by how well this newly dubbed "bio-energy" could flow, accumulate and neutralize imbalances within the body.

At about the time he was told to take a cruise, MC was sent information about a Dr. William Frederick Koch. MC was doing quite well and only got to this information a few years later. Dr. Koch had devised a therapy that triggered an energetic reaction in the body by using substances derived from brain and heart tissues. Dr. Koch had observed that brain and heart tissues were resistant to starvation. He had also noted that cancer as well as other diseases resulted from the breakdown in the body's oxidation. Dr. Koch concluded that there was a substance in these two tissues that could produce "energy." He could later synthesize these carbonyls from chemicals and produce Glyoxylide and Malonide.

Dr. Koch's Glyoxylide therapy included a strict diet of raw vegetables and vitamins as well as enemas. MC saw a striking resemblance to Dr. Heil Mittel Mandel's nutritional enzyme therapy. Along with a vigorous cleansing and detoxifying protocol, Dr. Mandel gave his patients supplements from animal, (mainly beef and pork), tissues such as heart, brain, liver, lymph and thyroid. An important difference was that Dr. Mandel went by William Kelley's findings that one diet did not fit all cases. Another important difference was that Dr. Koch injected these modified carbonyls intravenously as a vaccine, which was diluted in a homeopathic manner. It quickly energized the oxidation system manifold. It was true medicine, applying a small harmless dose to create a large healthy response.

In a very short time, it fired up the immune system, and one of the benefits was the elimination of cancer. Dr. Koch was persecuted, put on trial, and would eventually leave the country. MC, under Dr. Mandel, got his immune system to do the same thing more gradually through ingestion.

Since there seemed to be an organized effort to destroy any cure for cancer, Dr. Mandel limited his enzyme therapy to using only supplements that could be swallowed or absorbed through the skin. He refused to prescribe substances that would be outlawed in the country by the FDA. For example, it was against the law to prescribe laetrile (vitamin B17) or the use of apricot seeds, but it would be impossible to outlaw the consumption of twenty raw almonds per day, which furnished all the laetrile MC would need.

Dr. Mandel's main concerns were agencies trying to regulate the dosages of vitamins or enzymes. Just such a threat was the Dietary Supplement Safety Act of 2003. The act also called for regulating "complementary and alternative medicine products" and going as far as having certain "functional foods," that are recommended in treating disease, categorized and regulated as drugs.

The most intriguing thing to MC was how a food and drug agency could pursue 'energy users'. How could they interfere with the use of an energy, which they claimed did not exist?

MC remembered talking to Mr. Ross, the old caretaker at Reich's Orgone Energy Observatory in Rangeley, Maine. He said that he was ordered by the FDA to take apart all the orgone boxes, small and large. Mr. Ross was a carpenter. He neatly dismantled and sorted all the wood, sheet metal, steel wool and insulation in separate piles. He was flabbergasted when one of the FDA agents took a sharp tool and perforated all the insulation. The agents then threw out all the materials. The FDA played the role of decontaminators against an elusive bountiful energy. It was the visible world outlawing the non-visible world. Madame Curie's discoveries, on the other hand, were heralded as the probable future cure for cancer. Even though Madame Curie died of cancer caused by her exposure to radioactivity, the industry of radiation is given free rein to experiment on cancer patients. The lack of real breakthroughs in the field does not seem to faze government agencies. Yet, when other "energy" phenomena were applied to a successful cancer treatment, they were obliterated.

The first chemotherapies ever used were inspired by observations on the effects of poisonous gas on humans during World War I. Today, the National Cancer Institute (NCI) is synonymous with and is located at the old facility of the biological warfare research center at Fort Detrick, Maryland.

In the 20th Century, John Beard, W. Koch, W. Reich, Royal Rife and William Kelley, to name just a few, were defamed because of their insights about cancer.

MC had become aware of living in this ocean of life energy. It was difficult to talk about. Most people were not interested in knowing of it. It seemed to give MC energy to go on. He felt inspiration come from it. Sharinagar often said that the brain in its normal state was just a storehouse of memories. To heal, the brain needed to be silenced by this energy with the help of the lungs and the heart.

Dr. Koch's theory about the brain and heart tissue with these unusual properties came to mind. Sharinagar would further describe this energy as a presence that was always there. When aware of it, one could be healed, heal, be inspired and perhaps guided. It did not come of us but through us. It required a little willingness from the self-centered brain to enter into this presence.

MC had managed to find his cure for cancer. It was not necessary to understand what energy's role was - others had done so. Its many names did not help MC recognize what this energy was. Much like an electrician, who can manipulate electricity and use it without knowing what it really is, MC knew by now what enhanced energy's presence and what deterred or confounded it.

On occasion, MC, (without absolute proof), would sit back and give the energy permission to just do whatever it does; allow it to flow naturally without trying to control it -- embrace it without fighting or fearing it.

Among the problems on our planet, there are resolutions all about us.

The worldly-minded have gone on tangents, tackling health problems in an adversarial, confrontational and moralistic manner with vested interests in mind.

In times of chaos there is a need to become resolute. That time is with us, but so are the solutions.

REFERENCES

"One Man Alone: An Investigation of Nutrition, Cancer, and William Donald Kelley" by Nicholas J. Gonzalez, M.D. New Spring Press, 2010

"What Went Wrong: The Truth Behind the Clinical Trial of the Enzyme Treatment of Cancer" by Nicholas J. Gonzalez, M.D.

"The Trophoblast and the Origins of Cancer: One Solution to the Medical Enigma of Our Time" by Nicholas J. Gonzalez, M.D. & Linda L. Isaacs, M.D., 2009

"The Body Electric: Electromagnetism and the Foundation of Life", by Robert Becker & Gary Selden, 1998

"Politics in Healing: The Suppression and Manipulation of American Medicine" by Daniel Haley, 2000

"The Life and Trials of Gaston Naessens: The Galileo of the Microscope", by Christopher Bird, 1990

"The Cancer Cure that Worked! Fifty Years of Suppression" by Barry Lynes, 1987

"The Murder of Christ: The Emotional Plague of Mankind", by Wilhelm Reich, 1953

"The Oranur Experiment: First Report, 1947-1951" by Wilhelm Reich

REFERENCES (cont.)

"La Théorie Du Microzyma Et La Système Microbien" by Antoine Béchamp, 1920's

"Knowledge of the Higher Worlds and its Attainment: An Esoteric Spiritualism Initiation" by Rudolf Steiner

"Gifts from the Retreat" by Tara Singh, 1980

"A Course in Miracles", Foundation for Inner Peace, 1975

"The Miracle of Fasting: Proven Throughout History for Physical, Mental, & Spiritual Rejuvenation"by Paul C. Bragg & Patricia Bragg, 2004

"Knockout: Interviews with Doctors Who Are Curing Cancer--And How to Prevent Getting It in the First Place" by Suzanne Somers, 2010